Senescence in Plants

Senescence in Plants

Edited by **Brandon Chesser**

New York

Published by Callisto Reference,
106 Park Avenue, Suite 200,
New York, NY 10016, USA
www.callistoreference.com

Senescence in Plants
Edited by Brandon Chesser

International Standard Book Number: 978-1-63239-558-0 (Hardback)

Contents

Permissions

List of Contributors

Preface

This book explains the phenomena related to ageing in plants. It covers various aspects relating to ageing and discusses the factors associated with it. It provides elucidative information regarding ageing processes and senescent changes that living creatures such as plants face. Renowned experts and researchers in the field of senescence have made their valuable contributions to this book. It aims to benefit students and researchers interested in this field.

This book has been the outcome of endless efforts put in by authors and researchers on various issues and topics within the field. The book is a comprehensive collection of significant researches that are addressed in a variety of chapters. It will surely enhance the knowledge of the field among readers across the globe.

It is indeed an immense pleasure to thank our researchers and authors for their efforts to submit their piece of writing before the deadlines. Finally in the end, I would like to thank my family and colleagues who have been a great source of inspiration and support.

Editor

Plant

1

Some Aspects
of Leaf Senescence

Hafsi Miloud and Guendouz Ali
Laboratory of Improvement and Development of Livestock and
Crop Production, Department of Agronomy,
Faculty of Natural Sciences and Life,
Ferhat ABBAS University, Setif
Algeria

1. Introduction

The word *senescence* derives from two Latin words: *senex* and *senescere*. *Senex* means 'old'; this Latin root is shared by 'senile', 'senior', and even 'senate'. In ancient Rome the 'Senatus' was a 'council of elders' that was composed of the heads of patrician families. *Senescere* means 'to grow old'. The Merriam-Webster online dictionary defines *senescence* as 'the state of being old or the process of becoming old'. Aging is also the process of getting older. Therefore, aging has been regarded as a synonym of senescence, and the two words have often been used interchangeably, which, in some cases, is fine but in some other cases causes confusion. This paper will first briefly discuss the terminology of senescence, and then will review the literature related to mitotic senescence, a topic that has not been well discussed in the plant senescence research area and discuss some results relating to nutrient remobilization during leaf senescence.

2. Terminology and types of senescence

Senescence is a universal phenomenon in living organisms, and the word *senescence* has been used by scientists working on a variety of systems, such as yeast, fruit fly, worm, human being and plants. However, the meaning of the word *senescence* to scientists working on different organisms can be different, and the difference can be subtle in some cases and very obvious in some other cases.

3. Plants exhibit mitotic senescence, post mitotic senescence and cell quiescence

Plants exhibit both types of senescence. An example of mitotic senescence in plants is the arrest of apical meristem; the meristem consists of non differentiated, germ line-like cells that can divide finite times to produce cells that will be then differentiated to form new organs such as leaves and flowers. The arrest of apical meristem is also called proliferative senescence in plant literature .This is similar to replicative senescence in yeast and animal cells in culture. Another example of mitotic senescence is the arrest of mitotic cell division at early stages of fruit development. Fruit size is a function of cell number, cell size and intercellular space, and cell number is the major factor.

Cell number is determined at the very early stage of fruit development and remains unchanged thereafter. Post mitotic senescence occurs in some plant organs, such as leaves and floral petals. Once formed, cells in these organs rarely undergo cell division; their growth is mainly contributed by cell expansion; thus, their senescence, unlike mitotic senescence, is not due to an inability to divide. This type of senescence involving predominantly somatic tissues is very similar to that.

4. Physiological regulation

Reproductive development appears to play an important role in regulating proliferative senescence in plants, which is especially true in many monocarpic plants. Hensel *et al.* (1994) found that meristems of all inflorescence branches in the wild-type *Arabidopsis* ecotype Landsberg *erecta* (L*er*) ceased to produce flowers coordinately, but such a coordinated proliferative arrest did not occur in the wild-type L*er* plants with their fruits surgically removed. Similarly, meristem arrest was not observed in a male-sterile line that never sets seeds. This result suggests that the arrest of inflorescence meristems is regulated by developing fruits/seeds (Hensel *et al.*, 1994). Hensel *et al.* further proposed two models to explain the effect of developing fruits on the mitotic activity of meristems. One model is that a factor necessary for sustaining mitotic activity at the SAM is gradually taken and eventually depleted by developing fruits, resulting in arrest. The other model is that developing fruits produce a negative regulator of mitotic activities and that the negative regulator is transferred to and accumulated in the SAM to a threshold level so that the SAM is arrested. The factor, either positive or negative, is unknown.

5. Nutrient remobilization during leaf senescence

Senescence is the last stage in the development of leaves and other plant organs. While many plants are perennial (barring adverse conditions leading to premature death), and some species even very long-lived (at least from a human perspective), senescence and death of organs such as leaves is often an annual event. Due to its importance for agriculture, the senescence of annual crops (e.g. corn, rice, wheat, barley and some legumes) has been most intensely studied (Feller & Fischer, 1994; Hayati *et al.*, 1995; Crafts-Brandner *et al.*, 1998; Yang *et al.*, 2003; Robson *et al.*, 2004; Parrott *et al.*, 2005;Weng *et al.*, 2005). Additionally, as in other areas of plant science research,*Arabidopsis* has emerged as an important model system (Diaz *et al.*, 2005; Levey & Wingler, 2005; Otegui *et al.*, 2005). These plants show monocarpic senescence, i.e. fruit set and maturation are directly associated with whole-plant senescence and death. Other types of senescence, such as top senescence (in species with bulbs, tubers, tap roots or rhizomes), deciduous senescence (in some trees and shrubs of temperate climate zones) and progressive senescence (e.g. in evergreen trees) have received less attention. In contrast to annuals, leaf (or whole-shoot) senescence is often not directly associated with seed filling in perennial plants (Feller & Fischer 1994; Nood´en *et al.*, 2004). However, nutrient remobilization from senescing plant parts to surviving structures is a hallmark of the 'execution' of the senescence process in both annual plants, in which nutrients are retranslocated to the seeds, and perennial species, in which nutrients are transported to surviving structures such as bulbs and roots.

Plants need a number of elements in higher quantities or concentrations to complete their life cycle (macronutrients, including C, O, H, N, P, S, K, Mg and Ca), while a number of additional elements (micronutrients, including Fe, Mn, Zn, Cu, B, Mo, Cl and Ni) are needed in comparatively small quantities (Marschner, 1995). Some elements are essential only for specific taxonomic groups (e.g. Na, Si) and/or are considered beneficial (Marschner, 1995).

5.1 Nitrogen remobilization

Quantitatively, nitrogen is the most important mineral nutrient in plants (Marschner, 1995). It is often a limiting factor for plant growth, yield and/or quality (Gastal & Lemaire, 2002; Good et al., 2004). Additionally, as for carbon, the principal form in which many plants acquire nitrogen from the environment (nitrate) is more oxidized than the form in which it can be integrated into metabolites and macro molecules, demanding substantial energy input for the synthesis of nitrogen compounds. Although the biochemistry involved is different, the establishment and maintenance of a symbiosis with N2-fixing microorganisms (e.g. in legumes) is also costly (Crawford et al., 2000; Lodwig & Poole, 2003). For these reasons, efficient N remobilization increases the competitiveness of wild plants. Additionally, due to the economic and ecological (N runoff from agricultural soils) cost of N fertilization, this trait is of considerable importance to farmers.

In most plant tissues, the largest fraction of organic nitrogen, which is potentially available for remobilization during senescence, is contained in proteins. In photosynthetically active tissues of C3 species, over 50% of this nitrogen is found in soluble (Calvin cycle) and insoluble (thylakoid) chloroplast proteins (Peoples and Dalling, 1988; Feller and Fischer, 1994). Intriguingly, ribulose-1,5-bisphosphate carboxylase/ oxygenase (Rubisco) alone represents 50% of the total plastidial nitrogen.

All other cellular nitrogen fractions, including cytosolic and other proteins, nucleic acids, chlorophylls and free amino acids, while not negligible, represent relatively minor stores of organic nitrogen. Efforts at understanding nitrogen remobilization during leaf senescence have therefore focused on the biochemistry of plastidial protein degradation. Mae et al. (1983), using elegant 15 N-labeling techniques, have demonstrated that the synthesis and degradation phases of Rubisco are surprisingly clearly separated during leaf development. High rates of synthesis were observed until full leaf expansion; after this point, synthesis was minimal, but degradation rates started to increase. In this context, it is well known that the photosynthetic capacity of a leaf declines early during leaf senescence, while mitochondrial integrity and respiration are maintained longer (Gepstein, 1988; Feller and Fischer, 1994). That efficient N remobilization is associated with (early) loss of CO2 assimilation represents a formidable problem in annual crops. In this context, agronomists are well aware of the negative correlation between seed protein and yield.

5.2 Macro- and micronutrient remobilization

Developing (young) leaves constitute significant net importers ('sinks') for all nutrients, which are utilized to build the organ's cellular and molecular components. After the so-called sink–source transition (Ishimaru et al., 2004; Jeong et al., 2004), leaves become net exporters ('sources') of carbohydrates from photosynthesis, while import (through the xylem) and export (through the phloem) of phloem-mobile nutrients are (roughly) at an equilibrium in mature leaves (Marschner, 1995). The onset of leaf senescence is associated with a transition to net export of 'mobile' (see below) compounds, i.e. total (per leaf) content of some nutrients starts to decrease (Marschner, 1995). The literature often refers to this situation as 'redistribution', 'retranslocation', 'resorption' or 'remobilization' (Marschner, 1995; Killingbeck, 2004).

The main transport route from senescing leaves to nutrient sinks is the phloem (Atkins, 2000; Tilsner et al., 2005). Using various approaches, including sampling and analysis of phloem sap and (radioactive) tracer studies, it has been established that macronutrients with the exception of calcium (i.e. N, P, S, K and Mg) are generally highly mobile in the phloem, while micronutrients with the exception of manganese (i.e. Fe, Zn, Cu, B, Mo, Cl and Ni) show at least moderate mobility (Marschner, 1995). As a consequence, while some mobile nutrients decrease during leaf senescence, this is not true for calcium, which continues to accumulate throughout a leaf's life span. The molecular form, in which nutrients fulfill their biological functions, determines the biochemical steps necessary to make them phloem mobile. A certain percentage of many nutrients is biochemically inert, and cannot be remobilized (Marschner, 1995; Killingbeck, 2004). Cell wall components are a good example, and explain why fully senesced (dead) leaves are usually rich in carbon as compared to nitrogen. Some macronutrients, including carbon, nitrogen, phosphorus and sulfur, are covalently bound in myriads of both low-molecular-weight metabolites and macromolecules. Proteins and nucleic acids are important stores of nitrogen, phosphorus (nucleic acids) and sulfur (proteins); these macromolecules have to be degraded by specific hydrolases prior to phloem loading and transport. Metals (both macro- and micronutrients) can also be tightly bound, mostly by macromolecules, e.g. cell wall compounds or proteins. Their release is therefore often linked with the degradation of the functional complexes/macromolecules, to which they belong.

5.3 Carbon

Because it is taken up in gaseous form and a large amount of energy is needed for its reduction prior to its incorporation into metabolites, carbon occupies a special position in plant metabolism. Additionally, as discussed obove, degradation of the photosynthetic apparatus is an early event during leaf senescence, leading to a decrease of photoassimilate production and export to sinks, and to an increasing dependence of senescing tissues on respiratory metabolism (Gepstein, 1988; Feller & Fischer, 1994). Metabolization and, to some degree, remobilization of reduced carbon are therefore important for senescing leaves. In this context, Gut and Matile (1988, 1989) observed an induction of key enzymes of the glyoxylate cycle, isocitrate lyase and malate synthase, in senescent barley leaves. Based on these data, and based on low respiratory quotients (0.6), these authors suggested a reutilization of plastidial (thylakoid) lipids via β-oxidation, glyoxylate cycle and gluconeogenesis, allowing export of at least some of the carbon 'stored' in plastidial lipids from the senescing leaf. These observations have since been confirmed and extended (Pistelli *et al.*, 1991; Graham *et al.*, 1992; McLaughlin & Smith, 1994). He and Gan (2002) have shown an essential role for an *Arabidopsis* lipase in leaf senescence; however, it is not yet clear if this or other lipases are involved in preparing substrates (free fatty acids) for β-oxidation and gluconeogenesis. Roulin *et al.* (2002) have found an induction of $(1\rightarrow3, 1\rightarrow4)$-$\beta$-d-glucan hydrolases during dark-induced senescence of barley seedlings, suggesting a remobilization of cell wall glucans under these conditions.

Using radioactive labeling studies,Yang *et al.* (2003) demonstrated considerable remobilization of pre-fixed 14C from vegetative tissues to grains in senescent wheat plants. Interestingly, this process was enhanced under drought conditions, when leaf photosynthetic rates declined faster. Together, these data suggest that while C remobilization during leaf senescence has received less attention than N remobilization, it probably makes important contributions to seed development, at least in annual crops.

5.4 Sulfur

Besides carbon and nitrogen, sulfur is the third nutrient, which (relative to its main form of uptake, sulfate) is reduced by plants prior to its incorporation into certain metabolites and macromolecules. It is noteworthy, however, that plants also contain oxidized ('sulfated') sulfur metabolites (Crawford et al., 2000). Identically to carbon and nitrogen, sulfur is an essential element of both low-molecular weight compounds (including the protein amino acids cysteine and methionine) and macromolecules (proteins). Glutathione (γ -glutamyl-cysteinyl-glycine) represents the quantitatively most important reduced sulfur metabolite; it can reach millimolar concentrations in chloroplasts (Rennenberg and Lamoureux, 1990). Sulfur remobilization from older leaves has been shown; however, the extent of its retranslocation appears to depend on the nitrogen status, at least in some systems (Marschner, 1995). Sunarpi & Anderson (1997) demonstrated the remobilization of both soluble (non protein) and insoluble (protein) sulfur from senescing leaves. This study also indicated that homoglutathione (containing β-alanine instead of glycine) is the principal export form of metabolized protein sulfur from senescing soybean leaves.

5.5 Potassium

Next to nitrogen, potassium is the mineral nutrient required in the largest amount by plants. It is highly mobile within individual cells, within tissues and in long-distance transport via the xylem and phloem (Marschner, 1995). In contrast to the nutrients discussed above, potassium is not metabolized, and it forms only weak complexes, in which it is easily exchangeable. Next to the transport of carbohydrates and nitrogen compounds, potassium transport has been studied most intensely, using both physiological and molecular approaches (Kochian, 2000). Many plant genes encoding K+ transporters have been identified, and some of them have been studied in detail in heterologous systems, such as K+-transport-deficient yeast mutants. Similarly to the situation discussed for nitrogen transport, analysis of K+ transport is complicated by the fact that these transporters are organized in multigene families with (partially?) redundant functions (Kochian, 2000). Potassium was repeatedly reported to be remobilized in significant quantities from senescing tissues (Hill et al., 1979; Scott et al., 1992; Tyler, 2005). However, it has to be considered that this element easily leaches from tissues, especially senescing tissues (Tukey, 1970; Debrunner & Feller, 1995). Therefore, actually remobilized potassium quantities may be smaller than those reported in the literature.

5.6 Phosphorus

Unlike carbon dioxide, nitrate and sulfate, phosphate (main form of P uptake) is not reduced, but utilized in its oxidized form by plants (Marschner, 1995), both in lowmolecular- weight metabolites and in macromolecules (nucleic acids). Studies on P remobilization from senescing leaves are scarce. Snapp and Lynch (1996) concluded that in maturing common bean plants, leaf P remobilization supplied more than half of the pod plus seed phosphorus. In contrast, Crafts-Brandner (1992) observed no net leaf P remobilization during reproductive growth of soybeans cultivated at three different P regimes. Therefore, while P is a mobile nutrient, its remobilization may be influenced by a number of exogenous and endogenous/genetic factors, making generalizations on the importance of its remobilization difficult. Nucleic acids (especially RNA) constitute a major phosphorus store but, depending on the species and growth condition investigated, considerable P amounts are also present in lipids, in esterified (organic) form, and as inorganic phosphate (Hart & Jessop, 1984; Valenzuela et al., 1996). Similarly to the situation with nitrogen 'bound' in proteins, release of phosphorus from nucleic acids depends on the activities of hydrolytic enzymes. A decrease in nucleic acid levels is typical for senescing tissues, and increases in nuclease activities have also been observed (Feller and Fischer, 1994; Lers et al., 2001), indicating that if P is remobilized from senescing tissues, at least part of it is derived from the degradation of RNA and DNA.

5.7 Magnesium, calcium and micronutrients

Magnesium has not often been considered in studies on nutrient remobilization. However, despite the fact that this element is considered phloem mobile (Marschner, 1995), available results indicate a tendency of continued accumulation during leaf senescence (Killingbeck, 2004). Unsurprisingly, calcium, which is the least mobile of all macronutrients (Marschner, 1995), has repeatedly been found to increase in senescing leaves (Killingbeck, 2004).

Information on remobilization of micronutrients does not allow a generalized picture. For several of them, including Fe, Cu, Mn (which is the least phloem mobile among the micronutrients) and Zn, both remobilization from and accumulation in senescing leaves have been reported (Killingbeck, 2004, and references cited therein). Tyler (2005) gives a broad overview of the fate of numerous elements (including the micronutrients Fe, B, Mn, Zn, Cu, Mo and Ni) during senescence and decomposition of Fagus sylvatica leaves; however, in view of the results cited above, it is probably not possible to generalize conclusions from this study, e.g. with regard to the situation in annual crops.

5. Conclusions

This paper discussed some results relating to nutrient remobilization during leaf senescence.complex regulatory network controlling senescence in plants may be the result of selection pressure driven by different environmental stresses for the development of senescence.focus on limited number of model plant systems studied by plant senescence scientists may be required for more efficient research, and is likely to be highly relevant to agriculture as well as to our basic understanding of the senescence process in plants.

7. References

1] Crafts-Brandner, S.J. (1992).Phosphorus nutrition influence on leaf senescence in soybean. Plant Physiol 98, 1128-1132.

2] Crafts-Brandner, S.J., Holzer, R. & Feller, U. (1998).Influence of nitrogen deficiency on senescence and the amounts of RNA and proteins in wheat leaves. Physiol Plantarum 102,192-200.

3] Crawford, N.M., Kahn, M.L., Leustek, T. & Long, S.R. (2000). Nitrogen and sulfur in Biochemistry and Molecular Biology of Plants (Eds Buchanan, B., Gruissem, W.and Jones, R.).American Society of Plant Physiologists, Rockville, MD, pp.786-849.

4] Diaz, C., Purdy, S., Christ, A., Morot-Gaudry, J.-F., Wingler, A. & Masclaux Daubresse, C. (2005). Characterization of markers to determine the extent and variability of leaf senescence in Arabidopsis. A metabolic profiling approach. Plant Physiol 138,898-908.

5] Debrunner, N. & Feller, U. (1995).Solute leakage from detached plant parts of winter wheat: Influence of maturation stage and incubation temperature. J Plant Physiol 145,257-260.

6] Feller, U. & Fischer, A. (1994). Nitrogen metabolism in senescing leaves. Crit Rev Plant Sci 13(3), 241-273.

7] Gastal, F. & Lemaire, G.(2002). N uptake and distribution in crops: an agronomical and ecophysio- Logical perspective. J Exp Bot 53(370), 789-799.

8] Gepstein, S. (1988).Photosynthesis. In: Senescence and Aging in Plants (edsNood' en, L.D.and Leopold, A.C.).Academic Press, SanDiego, CA, pp.85-109.

9] Good, A.G., Shrawat, A. K. & Muench, D.G. (2004).Can less yield more? Is reducing nutrient in put into the environment compatible with maintaining crop production? Trends Plant Sci 9 (12), 597-605.

10] Graham, I.A., Leaver, C.J. & Smiths. (1992).Induction of malate synthase gene expression in Senescent and detached organs of cucumber. Plant Cell 4,349-357.

11] Gut, H. & Matile, P. (1989).Break down of galactolipids in senescent barley leaves. Bot Acta 102, 31-36.

12] Gut, H.& Matile, P. (1988). Apparent induction of key enzymes of the glyoxylic acid cycle in senescent barley leaves. Planta 176,548-550.

[13] Hart, A. L. & Jessop, D.(1984).Leaf phosphorus fractionation and growth responses to phosphorus of the forage legumes *Trifolium repens*, *T.dubium* and *Lotus pedunculatus*. Physiol Plant 61, 435–440.

[14] Hayati, R., Egli, D.B. & Crafts-Brandner, S.J. (1995).Carbon and nitrogen supply during seed filling and leaf senescence in soybean. Crop Sci 35, 1063–1069.

[15] He, Y. & Gan, S. (2002).Agene encoding anacyl hydrolase is involved in leaf senescence in Arabidopsis. Plant Cell 14,805–815.

[16] Hensel, L.L., Nelson, M.A., Richmond, T.A. & Bleecker, A.B. (1994). The fate of inflorescence meristems is controlled by developing fruits in Arabidopsis. Plant Physiol 106,863–876.

[17] Hill, J., Robson, A.D.and Loneragan, J.F. (1979) .The effect of copper supply on the senescence and the retranslocation of nutrients of the oldest leaf of wheat. Ann Bot 44,279–287.

[18] Ishimaru, K., Kosone, M.,Sasaki, H.and Kashiwagi,T.(2004).Leaf contents differ depending on the position in a rice leaf sheath during sink–source transition. Plant Physiol Biochem 42,855– 860.

[19] Jeong,M.L.,Jiang,H.,Chen,H.-S.,Tsai,C.-J.andHarding,S.A.(2004)Metabolic profiling of the sink-to-source transition in developing leaves of quaking aspen. Plant Physiol 136, 3364–3375.

[20] Kilian, A., Stiff, C. & Kleinhofs,A.(1995) . Barley telomerees shorten during differentiation but grow in callus culture. Proc Natl Acad Sci U SA 92, 9555–9559.

[21] Killing beck, K.T. (2004) .Nutrient resorption.In: Plant Cell Death Processes (ed.Nood´ en, L.D.). Elsevier Academic Press, Amsterdam, pp.215–226.

[22] Kochian, L.V. (2000) .Molecular physiology of mineral nutrient acquisition, transport, and utilization. In: Biochemistry and Molecular Biology of Plants (Eds Buchanan, B., Gruissem, W. & Jones, R.).American Society of Plant Physiologists, Rockville, MD, pp.1204–1249.

[23] Lers, A., Lomaniec, E., Burd, S.& Khalchitski,A.(2001).The characterization of LeNUC1,a Nuclease associated with leaf senescence of tomato. Physiol Plantarum, 112,176– 182.

[24] Levey, S. & Wingler, A. (2005).Natural variation in the regulation of leaf senescence and relation to other traits in Arabidopsis. Plant Cell Environ 28,223–231.

[25] Lodwig , E.& Poole, P. (2003) Metabolism of Rhizobium bacteroids. Crit Rev Plant Sci 22, 37–78.

[26] Mae, T., Makino, A. & Ohira, K. (1983). Changes in the amounts of ribulose bisphosphate carboxylase synthesized and degraded during the life spanofrice leaf (Oryzasativa L.). Plant Cell Physiol 24(6), 1079–1086.

[27] Marschner, H.(1995) Mineral Nutrition of Higher Plants. Academic Press, London.

[28] Mc Laughlin, J.C. & Smith, S. M.(1994)Metabolic regulation of glyoxylate-cycle enzyme synthesis in detached cucumber cotyledons and protoplasts. Planta 195, 22–28.

[29] Nood´ en, L .D. Guiam, J.L. & John, I. (2004) .Whole plant senescence. In: Plant Cell Death Processes (Ed .Nood´ en, L.D.).Elsevier Academic Press, Amsterdam, pp .227–244.

[30] Otegui, M.S., Noh, Y.S., Martinez, D.E., et al. (2005)Senescence-associated vacuoles within tense Proteolytic activity develop in leaves of Arabidopsis and soybean. Plant J 41(6), 831–844.

[31] Parrott, D., Yang, L., Shama, L. & Fischer, A.M. (2005) .Senescence is accelerated, and several proteases are induced by carbon 'feast' conditions in barley (*Hordeum vulgare* L.) Leaves. Planta 222,989–1000.

[32] Peoples, M.B. & Dalling, M.J. (1988).The interplay between proteolysis and amino acid metabolism during senescence and nitrogen reallocation. In: Senescence and Aging in Plants (Eds Nooden, L.D. & Leopold, A.C.).Academic Press, San Diego, CA, pp.181–217.

[33] Pistelli, L., DeBellis, L. & Alpi, A. (1991) Peroxisomal enzyme activities in attached senescing leaves. Planta 184,151–153.

[34] Rennenberg, H. & Lamoureux, G.L. (1990). Physiological processes that modulate the concentration of glutathione in plant cells. In: Sulfur Nutrition and Sulfur Assimilation in Higher Plants (ed. Rennenberg, H.). XPB Academic PublishersB.V., The Hague, pp.53–65.

[35] Robson, P.R.H., Donnison, I.S. & Wang., et al. (2004).Leaf senescence is delayed in maize expressing the Agrobacterium IPT gene under the control of a novel maize senescence-enhanced promoter. Plant Biotechnol J 2,101–112.

[35] Roulin, S., Buchala, A.J.and Fincher, G.B. (2002). Induction of $(1{\rightarrow}3, 1{\rightarrow}4)$-β-D-glucan hydrolases in leave sofdark-incubated barley seedlings. Planta 215, 51–59.

[36] Scott, D.A., Proctor. & Thompson. (1992). Ecological studies on a low land ever green rainforest on Maraca island, Brazil.II: Litter and nutrient recycling. J Ecol 80,705–717.

[37] Snapp, S.S. & Lynch, J.P. (1996). Phosphorus distribution and remobilization in bean plants as influenced by phosphorus nutrition. Crop Sci 36,929–935.

[38] Sunarpi and Anderson, J.W. (1997) .Effect of nitrogen nutrition on remobilization of protein sulfur in the leaves of vegetative soybean and associated changes insoluble sulfur metabolites. Plant Physiol 115, 1671–1680.

[39] Tukey, H.B. (1970). The leaching of substances from plants. Annu Rev Plant Physiol 21,305–324.

[40] Tyler, G.(2005) . Changes in the concentrations of major, minor and rare-earth elements during leaf Senescence and decomposition in a *Fagus sylvatica* forest. Forest Ecol Manage 206,167–177.

[41] Yang,J.C.,Zhang,J.H.,Wang,Z.Q.,Zhu,Q.S.& Liu,L.J.(2003). Involvement of abscisic acid and Cytokinins in the senescence and remobilization of carbon reserves in wheat subjected to water Stress during grain filling. Plant Cell Environ 26,1621–1631.

[42] Valenzuela, J.L., Ruiz, J.M., Belakbir, A. & Romero, L. (1996). Effects of nitrogen, phosphorus and potassium treatments on phosphorus fractions in melon plants. Commun Soil Sci Plant Anal 27(5–8), 1417–1425.

[43] Weng, X.-Y., Xu, H.-X. & Jiang, D.-A. (2005). Characteristics of gas exchange, chlorophyll fluorescence and expression of key enzymes in photosynthesis during leaf senescence in rice plants. J Integr Plant Biol 47(5), 560–566.

2

Advances in Plant Senescence

Kieron D. Edwards,
Matt Humphry and Juan Pablo Sanchez-Tamburrino
Advanced Technologies (Cambridge) Ltd.
UK

1. Introduction

Senescence is an integral component of a plant's lifecycle, which refers to changes that take place as the plant matures. A general distinction between plant senescence and animal senescence is the events observed in the animal kingdom typically steer growth while plant senescence orchestrates a massive shutdown or coordinated cell death in response to various stimuli designed to facilitate survival of the plant itself or the plant species.

In order to assist in the survival of the plant species, a sequence of tightly regulated genetic events efficiently governs a plant's death. These events are observable in a variety of plant models and in the different plant parts such as leaves, petals, reproductive organs (stamens and style), root cap, cortex and germinating seed. Leaf senescence will be the primary focus of this chapter.

A popular aspect of leaf senescence is the bright hues that can be observed on trees and plants during Autumn. The brilliant burst of colour that precedes the browning of leaves is an indication of active metabolic changes that result in the recycling or redistribution of nutrients to other parts of the plant. Evidence indicates the primary purpose of senescence in plants is for mobilization and recycling, a phenomenon that has tremendous implications for crop growth and food production.

Senescence marks the final phase of a leaf's development thereby launching degradation processes integral to the recycling and redistribution of the leaf's nutrients. Plant growth regulators, reproduction, cellular differentiation and hormone levels are internal factors that influence senescence (Thomas and Stoddart 1980; Smart 1994). Environmental stress also influences growth and can promote premature senescence. Certain parts of the plant may be sacrificed to enhance the chances of survival for the rest of the plant. Environmental cues include stress factors that adversely affect plant development and productivity; such as: drought, waterlogging, high or low solar radiation, extreme temperatures, ozone and other air pollutants, excessive soil salinity and inadequate mineral nutrition in the soil (Thomas and Stoddart 1980; Smart 1994). These environmental cues may accelerate leaf senescence by affecting the endogenous factors previously mentioned (Alegre and Munné-Bosch, 2004). Regardless of the trigger, the endogenous and exogenous signals that induce senescence appear to be coordinated through a common signalling network (Hopkins, 2007) involving the signalling molecules ethylene, jasmonic acid (JA), salicylic acid (SA) and Abscisic Acid (ABA) (Smart, 1994; Buchanan-Wollaston et al., 2005; van der Graaff et al., 2006)

2. Progress of senescence in plants

The general purpose of a leaf is to gather and generate nutrients for the plant. As a green leaf grows and develops, it creates an organ packed with nutrients. When the plant no longer requires the leaf, the senescence process is induced and recycling of all the nutrients that can be remobilized occurs. Leaf death is the final stage in the process; however, death is actively delayed until all nutrients have been removed.

The dismantling of the leaf begins with the chloroplasts, the energy-generating, photosynthetic powerhouse of the plant. Not unlike another energy generating organelle (ie, the mitochondria), chloroplasts are semiautonomous and they possess their own genome with its inherent transcriptional and translational machinery. Gradually the chloroplasts shrink and transform into gerontoplasts, an artefact characterised by the disintegration of the thylakoid membranes and accumulation of the plastoglobulin (Friedrich and Huffaker, 1980; Mae et al., 1984). The process of breaking down the chlorophyll is so pronounced that chlorophyll loss and the associated yellowing of the leaves are commonly used as indicators of plant senescence (Noodén et al., 1997). Control of the process is so tightly regulated that experiments demonstrating the reversibility of senescence have shown that the chloroplasts can recover structural features, re-synthesize chloroplast proteins and re-commence photosynthesis (Thomas and Donnison, 2000; Zavaleta-Mancera et al., 1999).

Degradation and remobilization of the chloroplast proteins and RNA contribute nitrogen and other nutrients for seed growth (Wittenbach, 1978). The mechanisms governing degradation of the chloroplast are not completely understood and there are competing theories about where proteins are degraded; for example they may be degraded locally within the chloroplast or in a centralized vacuole for degradation (Hortensteiner and Feller, 2002). Findings that support the possibility that the photosynthetic machinery is degraded *in situ* by the chloroplast include the presence of chloroplast enzymes, which are localized hydrolases that catalyze the initial steps of chlorophyll breakdown. (Hortensteiner, 2006). Proteases of the Clp, FtsH and DegP families are also expressed in chloroplasts and representative genes for these proteases are up-regulated in senescing leaves (Sokolenko et al., 1998; Nakabayashi et al., 1999; Itzhaki et al., 1998; Haussühl et al., 2001). Despite this observation, chloroplastic proteases are unlikely to account for the degradation of most photosynthetic proteins (eg, Rubisco) during senescence. Senescence-associated vacuoles, with strong proteolytic activity, have been identified in senescing tissue and likely also contribute towards the degradation of soluble photosynthetic proteins (Hensell et al., 1993; Comai et al., 1989).

Chloroplast degradation is followed by lipid, protein and nucleic acid degradation. Membrane integrity and cellular compartmentalisation are maintained until the latter stages of leaf senescence (Lohman et al., 1994; Smart, 1994; Pruzinska et al., 2005). A decline in photosynthesis during senescence may result in sugar starvation leading to the activation of conversion of lipids to sugars. Thylakoid breakdown leads to release of lipids, which are known to be converted to sugars through the glycoxylate cycle (Buchanan-Wollaston and Ainsworth, 1997; Kim and Smith, 1994). The sugars produced by conversion of large amounts of lipids may be in excess to that required for respiration of the senescing leaves and this excess may be exported to other growing and demanding parts of the plant. It appears that the expression of genes for the enzymes participating in the process of gluconeogenesis for production of sucrose play an important role during senescence as the

;enes responsible for synthesis of the enzymes involved in gluconeogenesis are reported to
•e significantly expressed during this time (Buchanan-Wollaston and Ainsworth 1997; Kim
ınd Smith, 1994).

_eaf senescence also results in the breakdown of nucleic acids to purines and pyrimidines,
√hich ultimately degrade to small and transportable carbon and nitrogenous compounds
hat are transported to growing parts of the plant (Buchanan-Wollaston and Ainsworth,
1997). In addition to mobilization of carbon and nitrogen, other nutrients like sulphur and
netallic ions are also known to be transported from senescing leaves. Sugar content can also
•e modified at the onset of senescence. Generally crops under a limited nitrogen nutrition
ınd high light regimen undergo early senescence and this is usually accompanied by an
ncremental rise of sugar levels in the leaves. Sugar has been suggested to trigger a
;enescence response based on gain or loss function experiments with hexokinase genes,
ɔrincipal regulators of a glucose signalling pathway (van Doorn, 2008).

3. Regulation of senescence and potential for biotechnology

Before the advent of modern biotechnology, which enabled scientists to commence
deciphering the relationship between genes and life, senescence was perceived as an
uncoordinated collection of events resulting in the metabolic and physiological changes to
plant organs described above. The study of plant genetics, genomics, proteomics and more
recently metabolomics have altered this perception and demonstrated that the process is
dynamic and well organised. Below, a few examples are provided to emphasize the
importance of a better understanding of plant senescence and the consequent potential of
applications derived from that understanding.

Several techniques and different plant models have been employed in the pursuit of
understanding the genetic mechanisms underlying the changes in gene expression
associated with senescence. The process of senescence is initiated in source tissues
prompting dramatic changes in gene expression, during which genes involved in basic
metabolism, including photosynthesis and protein biosynthesis, are down-regulated while
those involved in programmed cell death and stress response and/or encoding various
hydrolytic enzymes are up-regulated (Hopkins et al., 2007; Lim et al., 2007). Not
surprisingly the initial discoveries involving Senescence Associated Genes (SAGs) were
made in the model plant Arabidopsis (Arabidopsis thaliana) by methods including differential
display (Lohman et al. 1994), senescence-specific enhancer trap line screening using a range
of senescence promoting factors (He et al., 2001), subtractive hybridization (Gepstein et al.,
2003) and microarray experiments (Andersson et al., 2004). Many of the genes expressed
during senescence of tissues encode hydrolytic enzymes that are capable of disassembling
the ultra-structure of the cell and the breakdown of macromolecules (Smart, 1994; Griffith et
al., 1997; Watanabe et al., 1994). In addition, a large number of transcription factors, as well
as genes encoding carbohydrate and nitrogen-mobilising enzymes, nucleases and stress-
responsive proteins, have been found to exhibit increased expression in senescing leaves
(Buchanan-Wollaston and Ainsworth, 1997; Comai et al., 1989; Kim and Smith, 1994). The
gene expression changes and biological processes that are up- and down-regulated during
senescence as indicated by such studies, have been reviewed elsewhere (Guo and Gan,
2005), so will only be touched on in this review. What is more in the scope of this review are
the potential implications that senescence has for plant biotechnological applications.

In addition to the conventional use of crops as food sources, innovations continue to expand the role of crop species in society. With these changes, the importance of understanding senescence becomes even more significant. Crops and trees are being developed as an alternative fuel source. Plants are also being integrated into the production of pharmaceutical ingredients and complex protein therapies such as vaccines (Lossl and Waheed, 2011). These and other innovative uses for plants make obtaining a greater understanding of senescence a necessary step for harnessing the influence of senescence on the plant lifecycle and reducing the impact this has on product yields and stability. The SAGs found through Arabidopsis investigations have provided a reference point for studies in other plant species, providing the potential to translate fundamental understanding into applied tools. Delaying the onset of senescence could increase the production of the desired plant product. This may be of particular interest in plastid expression systems (reviewed in Day and Goldschmidt., 2010), given that the chloroplast degradation occurs at a relatively early stage of senescence.

3.1 *Populus tremula* and bio fuel

As suggested above, crops are being developed for alternative applications including bio fuels and paper production. The deciduous Aspen tree species *Populus tremula* is one such plant being developed for alternative fuel production. By comparing expressed sequence tag (EST) libraries generated from young fully-expanded leaves to leaves harvested immediately prior to visible signs of senescence, Bhalerao et al., (2003) identified *P. tremula* homologs for many known Arabidopsis SAGs. Altering the expression of these SAGs may have an effect on dormancy in this species with possible implications on the wood yield from these trees.

The onset of growth cessation and dormancy represents a critical ecological and evolutionary trade-off between survival and growth in most forest trees. Without this dormant stage nutrients stored in green leaves would be lost to frost, which would impact growth in the spring. Tight regulation over the timing of senescence is thus important. Latitudinal clines influence the critical photoperiod for onset of bud set (dormancy) and leaf senescence in Aspen (Fracheboud et al., 2009). This cline in dormancy was associated with multiple alleles of PHYTOCHROME B2 (PHYB2), a photoreceptor that is related to light perception and light input to the circadian clock (the internal timing mechanism of the plant). The circadian clock enables the plant to co-ordinate its endogenous activities with the external environment to maximise the effectiveness of its activity. These activities occur on a daily basis, such as the timing of photosynthetic gene expression (Harmer et al., 2000, Edwards et al., 2006), and on an annual basis when measuring photoperiod and co-ordinating activities such as transitions to flowering, senescence or dormancy (reviewed in Jackson, 2009). Indeed, previous experiments suggest more accurate timing by the clock in relation to the external environmental cycles also has the potential to improve crop yields (Dodd et al., 2005). Such regulation governing the timing of critical events is relevant to all crop species grown in temperate climates. In the case of senescence, utilising regulatory mechanisms such as the circadian clock has the potential to alter the timing of this process with benefits to both wood production (reduced loss of nutrients to frost) as well as, for example, increasing the length of the grain filling period in other crops.

1.2 Impact on yield

During whole plant senescence, fixed carbon and nitrogen are mobilized to reproductive or storage organs, which are harvested for human consumption (Vierstra, 1996; Hopkins et al., 2007; Lim et al., 2007). The process of senescence impacts all crop species and so the increased understanding of the tight regulatory mechanisms that control the process could potentially have an immeasurable impact on the world's agricultural production. Whole plant senescence plays a key role in remobilizing and transferring nutrients into the vegetative tissue and eventually to grain. The grain filling period is a critical period because many processes can influence the final grain yield (Yang and Zhang, 2006). For example, delaying whole plant senescence can be achieved by heavy use of fertilizer or development of a stay-green phenotype produced using a genetic or transgenic strategy. Extending or delaying senescence is believed to augment the grain filling stage thereby increasing grain yield. Contrarily, stresses, such as drought, induce early senescence, prompting the reduction of photosynthesis and shortening the grain filling period (Gregerson et al., 2008) and thus having the opposite affect on yield. Ectopic expression of SAG101, a protein with acyl hydrolase activity, has been shown to cause precocious senescence in both attached and detached leaves of transgenic Arabidopsis plants (He and Gan, 2002). Antisense expression of the gene, resulting in repression of the endogenous genes expression, was shown to cause a delay in the onset of senescence (He and Gan, 2002). Utilising genes such as SAG 101 to induce a stay-green/delayed senescence phenotype could potentially be employed in biotechnological strategies to increase yields in crops such as wheat.

Effective recycling of nutrients could have a massive impact on crop yields. Recycling of carbon and nitrogen during senescence involves the sequestering of cytoplasm and organelles into special autophagic vesicles. These vesicles deliver their contents to the vacuole (or lysosome) for breakdown by localized hydrolases (Thompson and Vierstra, 2005; Bassham, 2007). The breakdown products are either consumed by the host cell or transported to other tissues and organs. Under normal growth conditions, autophagy takes place at a basal level. The process ramps up in response to nutritional demand, biotic or abiotic stresses, and senescence. Autophagy plays an important role in the proper recycling of nutrients especially as a plant scavenges available nutrients from storage tissues and older senescing leaves.

When a pathway has been highly conserved evolutionarily, other organisms can provide the reference point for understanding a system in plants. The genes associated with autophagy discovered in yeast, enabled investigators to identify homologous genes in Arabidopsis and, subsequently, in rice and maize. Genome searches of Arabidopsis identified a collection of proteins structurally and functionally related to many of the ATG components present in yeast (Thompson and Vierstra, 2005; Bassham, 2007). In an effort to determine the importance of autophagy to crop plants, investigators at the University of Wisconsin, using the Arabidopsis as a reference, described a collection of components that participate in the ATG8/12 conjugation cascades in both rice (*Oryza sativa*) and maize (*Zea mays*). Remarkably, all components required for ATG8/12 conjugation in yeast and Arabidopsis (Ohsumi, 2001; Thompson and Vierstra, 2005) were identified in both rice and maize suggesting that the pathway is highly conserved. The group went on to greater characterize the expression of the maize ATG genes (Chung et al., 2009). The investigators observed an increase in ATG transcripts during leaf senescence and under nitrogen and fixed-carbon limiting conditions. The results indicate that the highly conserved process of autophagy plays a key role in nutrient remobilization with some variations unique to maize. The description of the maize ATG system provides a set of molecular and biochemical tools to study autophagy in this crop under field conditions (Chung et al., 2009). The same is true for rice (Ohsumi, 2001; Thompson and Vierstra, 2005). This type of knowledge may help to reveal important control points in autophagy that could be manipulated in both food and bio fuel crops to enhance nutrient use efficiency or to better allocate carbon and nitrogen to specific organs for improved yield.

In addition to highly conserved genes, specific genes or gene families that can also be employed to influence grain quality and yield. Uauy et. al., (2006) cloned a Quantitative Trait Locus (QTL) associated with increased grain protein, zinc, and iron content known as Gpc-B1. The ancestral wild wheat allele encodes a functional NAC transcription factor (NAM-B1) that accelerates senescence and increases nutrient remobilization from leaves to developing grains. In contrast, modern wheat varieties carry a non-functional NAM-B1 allele. Reduction in RNA levels of the multiple NAM homologues by RNA interference delayed senescence by more than three weeks and reduced wheat grain protein, zinc, and iron content by more than 30%. Other examples of specific genes having an effect in senescence include the cytokinin synthesis gene IPT, which has been shown to delay leaf senescence (Gan and Amasino, 1995), thereby providing the potential to increase seed setting time and yield, but the affect this has on nutritional value must also be considered.

3.3 Ripening

Although the main focus of this review relates to leaf senescence, fruit ripening is an aspect of plant senescence that is also of global significance. The timing of ripening is a key consideration when harvesting and transporting fruit to market. Successful efforts to control fruit ripening are based on either reducing the biosynthesis of the plant hormone ethylene or slowing down the rate of fruit softening by targeting the genes involved in cell wall modification (Causier et al., 2002). The Flavr Savr tomato is an example of an early attempt to slow ripening using the latter strategy. Researchers at Calgene hoped to slow the ripening process of the tomato by engineering in an antisense gene to interfere with production of the enzyme polygalacturonase (Weasel, 2009). The enzyme normally degrades pectin in the cell walls and results in softening.

More recently, investigators have attempted to characterize the N-glycan processing enzymes and their role in during non-climacteric fruit softening. The plant hormone ethylene does not influence ripening of non climacteric fruits and different genes need to be targeted for the different categories of fruits (Causier et al., 2002). Two ripening-specific N-glycan processing enzymes, α-mannosidase (α-Man) and β-D-N-acetylhexosaminidase were identified in the fruit capsicum (*Capsicum annuum*, Ghosh et al., 2010). Using RNA interference to suppress production of such enzymes has the potential to improve the shelf life of fruits, with obvious implications for improved food stability/storage.

4. A view on tobacco biotechnology and senescence

Tobacco is different from many of the crops discussed above because the organ harvested for human use is the leaf rather than reproductive organs (i.e. seed and fruit). The smoke generated during the burning of tobacco is a complex mixture of thousands of chemicals (Rodgman and Perfetti, 2008). Research to identify and characterise the harmful components present in tobacco smoke is ongoing, with lists such as the 44 Hoffmann analytes, being produced by researchers and public health organisations (Hoffmann and Wynder 1967; Baker, 1999; Norman, 1999; Borgerding and Klus, 2005). A major focus of tobacco research is related to lowering the levels of these chemicals from smoke in an effort to reduce the harmful effects associated with tobacco use (for an overview of such research the scientific website of British American Tobacco [BAT] n.d.). Understanding the regulation and effect of senescence on tobacco leaf chemistry could be of particular importance to these traits since it is following the onset of senescence that 'ripe' tobacco leaves are harvested.

4.1 Gene expression changes in senescing tobacco leaves

The availability of microarrays has considerably increased the extent to which differentially expressed genes can be identified and this line of research provides valuable insights into the identification of senescence related genes. High throughput analysis makes it possible to monitor changes in gene expression throughout the lifecycle of a plant. Researchers from Advanced Technologies (Cambridge) have recently described the generation of a tobacco (*Nicotiana tabacum*) custom expression array (Edwards et al., 2010). This array was used to develop the Tobacco Expression Atlas (TobEA), a map of gene expression from multiple tissues sampled throughout the life cycle of the tobacco plant which can be used as a reference data set for plant researchers. The expression data is freely available via the Solanaceae Genomics Network (SGN), a web based genomic resource for plants of the Solanaceae family (Mueller et al., 2005). Studying the changes in gene expression has the potential to identify targets that enable modifications or changes to leaf constituents in tobacco using transgenic or non-transgenic (e.g. molecular breeding) approaches.

Included in the TobEA study was a set of leaves from different positions that were categorised into either green (sink) or four distinct senescent (source) leaves, based on the average amount of yellowing and chlorosis across the leaf (Figure 1A; data not shown). Analysis of the gene expression changes in the tobacco leaf series suggested that tobacco showed similar changes during the progression of senescence as Arabidopsis leaves. For example the defence-associated phytohormone SA is known to play a role in developmental leaf senescence in Arabidopsis, with mutants and transgenic lines defective in the SA-mediated signalling pathway exhibiting delayed senescence (Buchanan-Wollaston et al., 2005; Morris et al., 2000). A significant over-representation of genes associated with systemic acquired resistance and the SA-mediated signalling pathway was observed in the up-regulated genes in the TobEA dataset, including presumptive orthologues of *ENHANCED SUSCEPTIBILITY 1* (*EDS1*) and *PHYTOALEXIN DEFICIENT 4* (*PAD4*), central regulators of SA-mediated defence (Figures 1H and I; Feys et al., 2005; Morris et al., 2000).

There was also significant over-representation of Gene Ontology categories associated with defence against fungal pathogens as well as cell death and innate immune responses in leaves at more advanced stages of senescence. These included a homologue of the Arabidopsis basic chitinase *PR3* (Verburg and Huynh 1991) in addition to other components of plant immunity/defence. This supports the growing evidence that pathogen defence and senescence share common components (Quirino et al., 1999; Feys et al., 2005), presumably largely via the use of similar signalling pathways leading to accumulation of reactive oxygen species and cell death (Yoshida, 2003).

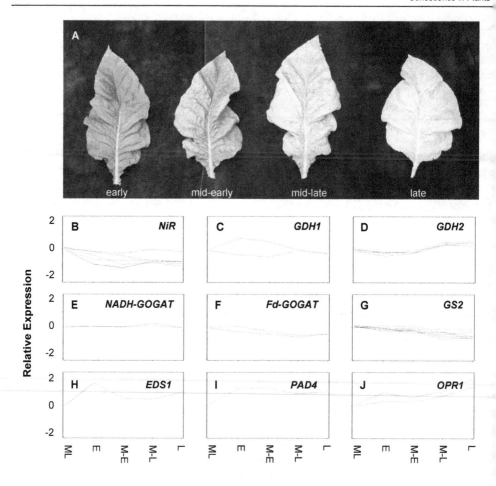

Leaf Type

Fig. 1. Gene expression changes in tobacco source and sink leaves
(A) A sink to source series of leaves harvested from different positions on tobacco plants included in the Tobacco Expression Atlas and categorised as early- (E), mid-early- (M-E), mid-late- (M-L) and late- (L) senescent leaves based on level of Chlorosis (TobEA; Edwards et al., 2010). (B-J) Log expression data for selected transcripts associated with Nitrogen metabolism and plant hormone responses shown (See top right of each plot for transcript identification). Expression data was pre-processed with RMA and normalised against mature leaf (ML) samples showing no visible signs of senescence (also included in the TobEA data set). Differentially expressed genes from the sink-source series versus the mature leaf control were identified by one-way analysis of variance with Tukey HSD post hoc testing in GeneSpring GX 10 (P < 0.05). Gene ontology analysis of the up- and down-regulated genes in each condition (described in main text) were analysed by a custom script as described previously (Edwards et. al., 2010).

In addition to SA, several other hormone pathways were identified as being over-represented in the TobEA leaf senescence dataset including Jasmonic acid (JA). JA is known to have a role in developmental senescence, with both levels of JA itself and a number of JA biosynthetic genes found to increase during senescence (He et al., 2002). A similar response was observed in the developmental senescence dataset, with 12-OXOPHYTODIENOATE REDUCTASE (OPR) family members being induced and an overall over-representation of genes involved in JA-meditated induced systemic resistance (Figure 1J; data not shown).

Ethylene is also known to play a role in promoting the onset of senescence (Grbic and Bleecker, 1995). Interestingly however, ethylene–mediated responses were not significantly over-represented in the TobEA data, suggesting that either ethylene is not as important in senescence of tobacco or that changes in this pathway are occurring post-transcriptionally. Stress can induce a senescence response in plants and one of the principal mediators of the stress response is the phyto-hormone Abscisic Acid or ABA (Smart 1994). ABA is a participant in drought (water) and cold stress responses (Wingler and Roitsch, 2008) and directly influences the sugar accumulation in response to stress. Interestingly, ABA will induce senescence during drought stress whereas it will delay senescence during cold stress (Xue –Xuan et al., 2010). In the TobEA data the ABA metabolic processes were significantly reduced late in senescence, the reason for which is unknown (data not shown).

Cytokinin levels in senescing leaves are though to play a key role in developmental leaf senescence, with both external and endogenous application resulting in delayed senescence (Smart 1994). This is largely reflected in the transcriptional responses to developmental senescence in Arabidopsis (Buchanan-Wollaston et al., 2005), as well as in the TobEA data, where cytokinin response processes were significantly down-regulated compared to controls.

Interestingly, phenylpropanoid biosynthesis was identified as a significantly down-regulated process the TobEA data. It would be expected that increased production of photo-protective phenylpropanoids, flavonoids in particular, would be observed during developmental leaf senescence, due to increased light stress during the degradation of chlorophyll (Buchanan-Wollaston 2005). Indeed, Buchanan-Wollaston et al., (2005) found a number of flavonoid biosynthesis genes had increased expression during developmental senescence.

Consistent with the phenotypic observations of the leaves themselves (Figure 1A), there was an enrichment of genes associated with photosynthesis being down-regulated in the TobEA data. This was accompanied by significant number of down-regulated genes associated with chloroplast components as well as responses to red, far-red and ultraviolet light stimuli. Over all the data suggest that similar processes occur during leaf senescence in Tobacco as in Arabidopsis and highlights the potential to translate findings in model species to biotechnological applications in other crops including Tobacco.

4.2 Nitrogen metabolism and harm reduction

Environmental and economic issues combined have increased the need to better understand the role and fate of nitrogen in crop production systems. Nitrogen is one of the most important nutrients recycled by the plant during senescence, with up to 90% recovered from the leaf during this process (reviewed in Liu et al., 2008). Adding nitrogen to the soil increases crop yields and delays senescence, whereas a reduced fertilizer regimen generally triggers early whole plant senescence in crops due to low nitrogen. A strong coordination of nitrogen-uptake, assimilation and remobilization is required for a beneficial grain filling stage (Hortensteiner and Feller 2002). The period that follows flowering can be critical in this process. Some crops, such as maize (C4 photosynthesis), use Nitrogen sourced both from the root's uptake and assimilation of NO_3- as well as nitrogen remobilized during leaf senescence. Other crops, such as oil seed rape, primarily rely on the remobilization of nitrogen from leaves, making these crops more dependent on the senescence process (Coque et. al., 2008). When nitrogen inputs to the soil system exceed crop needs, there is a possibility that excessive amounts of nitrate (NO_3-) may enter either ground or surface water causing a detrimental effect on the environment.

In the case of tobacco, a greater understanding of the metabolism of nitrogen could also be applicable in an attempt to reduce the harmful constituents contained in cigarettes. One class of chemicals likely to feature in any future legislation of the tobacco industry is the Tobacco Specific Nitrosamines (TSNAs); 4-(N-methlynitrosamino)-1-(3-pyridyl)-1-butanone (NNK), N-nitrosonornicotine (NNN), N-nitrosoanabasine (NAB) and N-nitrosoanatabine (NAT).

TSNAs are principally formed by the nitrosation of tobacco alkaloids during the curing (drying) of tobacco leaf (Burton et al., 1989; Burton et al., 1994; Hoffmann et al., 1994; Spiegelhalder and Bartsch, 1996). Several studies have demonstrated a significant correlation between nitrite (formed by the microbial reduction of nitrate during curing) and TSNA levels in tobacco leaf, leading to the proposal of nitrite as the key nitrosating agent for TSNA formation (Burton et al., 1989; Fischer et al., 1989; Burton et al., 1994; Spiegelhalder and Bartsch, 1996; Wu et al., 2005). Curing conditions (including airflow, temperature and humidity) and their affect on microbial activity have been shown to affect the levels of TSNAs formed (Burton et al., 1989; Burton et al., 1994). Further understanding the nitrogen metabolism of tobacco could aid in reducing the potential for accumulating nitrosating agents during the curing process helping to limit the formation of TSNAs, and potentially reducing the levels of these toxicants in tobacco smoke.

Figure 2 shows a summary of the nitrogen assimilation pathway in plants. Plant nitrogen assimilation primarily occurs in mesophyll cells and involves the reduction of nitrate (taken up by the root) into ammonia by the enzymes — Nitrate Reductase (NR) and Nitrite Reductase (NiR; Figure 2). The ammonia is subsequently assimilated into the amino acids glutamine (gln) and glutamate (glu) via the cyclic action of Glutamine Synthetase (GS) and Glutamine-2-oxoglutarate aminotransferase (GOGAT; Figure 2; Lea and Miflin 1974).

Nitrogen assimilation is regulated by many factors, including the availability of sugars and other metabolites and also shows significant variation over the diurnal cycle (reviewed in Stitt et. al., 2002). The expression and activity of genes involved in nitrogen reduction and assimilation have previously been shown to be down-regulated in tobacco leaves at more advanced stages of senescence (Mascluax et. al., 2000). Masclaux et. al, compared leaves from different positions on mature tobacco plants and showed that there was a switch between nitrogen assimilation and nitrogen recycling from sink to source leaves at more advanced stages of senescence. The leaf series included in the TobEA microarray data set described above is similar to the leaf series tested by Masclaux et. al (Masclaux et. al 2000). Expression of nitrogen metabolism genes in the TobEA leaves compared with fully expanded mature leaves showing no visible signs of senescence (also included in the TobEA data), demonstrated consistent results with Masclaux et al., (2000; Figure 1).

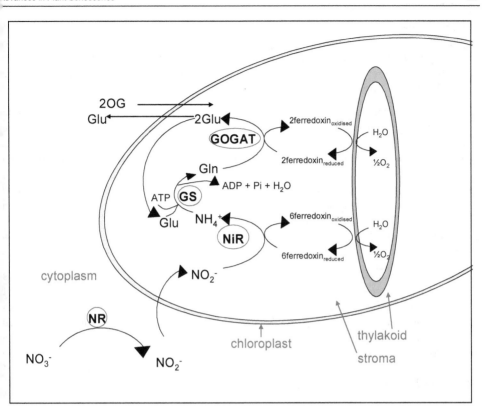

Fig. 2. Primary nitrogen assimilation in plants.
Representation of the primary nitrogen assimilation pathway in a mesophyll cell
(modified from Mohr and Schopfer, 1994), showing reduction of Nitrate to Nitrite and then
Ammonia by NR and NiR respectively and subsequent incorporation of nitrogen into gln
and glu by the cyclic activity of the enzymes GS and GOGAT.

No probe sets for NR could be identified on the tobacco array, so the reduction in expression
of this gene shown by Masclaux et al., (2000) could not be confirmed in the TobEA data.
However, a decrease in NiR expression over the leaf series was shown, supporting a
reduction in nitrogen fixation activity in the older leaves (Figure 1B). The ammonia
generated by NR and NiR activity is incorporated into amino acids by the GS-GOGAT cycle.
Plants have two types of GOGAT; ferredoxin dependent (Fd-GOGAT) and NADH
dependent (NADH-GOGAT). Similarly GS genes can be subdivided into cytosolic and
plastidic forms (GS1 and GS2 respectively). Fd-GOGAT functions in concert with GS2 and
NADH-GOGAT is associated with GS1. In previous studies (Buchanan Wollaston 2005, Lin
and Wu 2004), GS1 and NADH-GOGAT have demonstrated a co-ordinated increase in
expression in Arabidopsis. GS1 expression was previously shown to be induced in tobacco
source leaves, whereas GS2 transcripts were shown to be down regulated (Masclaux et al.,
2000). No tobacco orthologues for GS1 were identified on the tobacco microarray and
transcripts for NADH-GOGAT did not demonstrate changes in expression over the TobEA

dataset (Figure 1D). However, consistent with previous results, *GS2* transcripts were down regulated over the series of leaves (Figure 1C). A similar pattern of expression was also shown by tobacco *Fd-GOGAT* transcripts, supporting the proposed coordinated regulation and activity for these genes and a reduction of chloroplastic nitrogen assimilation in source leaves (Figures 1C and E).

Glutamate dehydrogenase (GDH) catalyses a reversible reaction adding or removing amino groups from glutamate. It has been proposed that the principal role of GDH is the deamination of glutamate in order to maintain a homeostatic balance of this amino acid that is thought to play a key role in the cross talk between the carbon and nitrogen assimilation pathways (Labboun et al., 2009). It has also been suggested that GDH amination may play a role in replacing glutamine synthetase (GS) activity in nitrogen assimilation within source leaves, which is lost during senescence (Masclaux et al., 2000). Previous studies have shown an increase in *GDH* expression in source leaves (Masclaux et al., 2000). Consistent with this, tobacco *GDH2* orthologs did show an increase in expression over the series; however, little change was shown by *GDH1* (Figures 1F and G).

Changes in the expression of genes involved in nitrogen assimilation shown by the leaves suggested that nitrogen metabolism was altered in source leaves towards remobilisation of the nitrogen resources to sink leaves. Consistent with this, gene ontology analysis of clusters of genes showing up-regulation in leaves demonstrating more advanced senescence revealed over representation of genes with functions related to proteolysis, the proteosome and endoplasmic reticulum associated protein catabolism. Increased understanding of the regulation of senescence in tobacco leaves could potentially help to limit the content of nitrate (and other nitrosonating agents) in harvested leaves prior to curing. This may augment efforts to reduce the levels of TSNAs in tobacco smoke; however, the study of senescence also provides other tools to facilitate TSNA reduction.

The main focus of agricultural research has been towards increased yield along with other agronomic traits. It is apparent in some crops that this has led towards an ignorance of flavour and texture components (as well as the associated nutritional value). Research is currently ongoing to understand and ultimately adjust the metabolic content of crops, such as those found in tomato, that contribute towards flavour and nutrition. This orientation toward flavour highlights the realisation that a perceived consumer benefit and consumer acceptance is becoming a more important driver in the development of new crops (Klee 2010). As indicated above, tobacco crops can be cured by multiple methods and the resulting leaf, or grades are blended together to produce the constituents of a cigarette. Dependent on the design, air cured tobacco typically only constitutes up to 30% of the blend in a cigarette, the rest being mainly made up of flue cured leaf (see Davis and Nielsen 1999 for a description of tobacco agronomy and chemistry). The conditions during air curing can lead to the formation of high levels of TSNAs. Thus, removing such grades from the blend could have beneficial affect on the overall level of TSNAs in the product. However, air cured grades make a significant contribution to the overall flavour of the cigarette, so the resulting product may not be consumer relevant and thus have no impact on harm reduction efforts. Replacing air cured grades with other grades that replace the flavour characteristics of these grades, but without the inherent higher levels of TSNAs provides one potential solution.

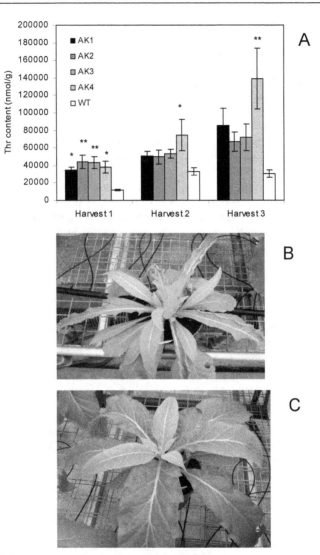

Fig. 3. Increased threonine production in tobacco leaves
Levels of threonine (nano moles per gram of cured leaf) for wild type tobacco and four
independent transgenic tobacco lines expressing a mutated form of the Arabidopsis
Aspartate Kinase AK:HSD (See inset key for line identification). Bars show mean threonine
levels and error bars represent Standard Error of the mean. Leaves were taken from three
harvest positions from the bottom to the top of the plant (Harvest 1 – 3). Asterisks represent
significant difference between transgenic lines and WT for each harvest position based on
one way analysis of variance with Tukey HSD post hoc testing (* $P < 0.005$, ** $P < 0.001$).
Constitutive expression of the same gene results in increased threonine levels in tobacco
leaves (data not shown), but results in reduced growth and altered morphology in (B)
transgenic plants compared to (C) wild type plants.

Amongst other differences, air cured tobaccos tend to have lower levels of sugars and an altered balance of free amino acids compared to flue cured tobacco leaf (Davis and Nielsen 1999). Threonine (thr) is one of the amino acids observed in higher levels in air cured leaf compared to flue cured, indicating it may contribute towards the flavour and aroma of the tobacco. Within the leaf, the biosynthetic pathway leading to production of thr is tightly regulated by a negative feedback control loop. In the case of feedback inhibition the end-product, in this case thr, competitively inhibits the activity of the bifunctional enzyme ASPARTATE KINASE (AK) (EC 2.7.2.4) -HOMOSEREINE DESATURATE (HSD) (EC 1.1.1.3) – and consequently blocks the enzymatic processes leading to its own synthesis (Shaul and Galili,1993). Disabling the enzyme that switches off thr production would prompt a greater accumulation of the compound, but, if the accumulation takes place too early in the plant's life cycle, the fitness of the plant is severely compromised (Fig 3A: data not shown). To overcome this obstacle, the promoter of the senescence associated gene SAG12, identified in Arabidopsis (Lohman 1984), was used to drive expression of mutated forms of AK:HSD gene from Arabidopsis in tobacco. Elevated leaf thr levels were achieved in the modified plants without compromising the plant's fitness (Fig 3 B and C). If the increase in thr results in an increase of the air cured 'flavour' in the tobacco, then such an approach could provide the potential to reduce the amount of this tobacco in the blend resulting in an associated reduction in the TSNA levels.

Optimising the timing and absolute level of expression by selecting other senescence associated promoters from tobacco could help to increase the yield of thr present in harvested leaves. Such promoters, could also be used to up, or down regulate the synthesis of other target flavour or toxicant precursors at the correct stage in the plant's life cycle to maximise the target phenotype, with limited effect on the growth of the plant.

5. The translational nature of innovation

Nature has long been an infinite resource for the purpose of scientific discovery. Consequently, the study of plant senescence possesses immeasurable potential for increasing the understanding of the plant kingdom and the technological application of that knowledge. Genomics and the subsequent disciplines of proteomics and metabolomics have provided a complete reorientation toward the ways in which plants are designed to facilitate preservation of their own species. Thousands of genes that increase expression during leaf senescence have been isolated from a number of crop varieties; such as: Arabidopsis, wheat, tomato, maize, rice, and tobacco; these are just the tip of the iceberg.

Many of the recent advances in the understanding of plants (and organisms in general) can be attributed to the exponential increase in the sequencing and bioinformatics capacity of the world's research communities, coupled with numerous initiatives being driven by governments, academics and the private sector. Advances in gene sequencing techniques have made it possible to decipher entire genomes and high throughput microarray analysis and other techniques make it possible to monitor changes in a plant over time. By comparing what genes are switched on and off as a plant senesces, a collection of SAGs have already been discovered. Tracing the homology of conserved sequences through the evolutionary line, not only has facilitated the discovery of more SAGs, it has helped to elucidate the dynamics of an entire senescence-associated biosynthetic pathways such as ATG8/12. Comparative studies between species not only reveal similarities, researchers inevitably find unique differences specific to the plant variety and species contributing even more information to the pool.

Alternatively, Biotechnological tools (i.e. gene vectors) and progressive strategies, such as molecular breeding, make it possible to apply research findings to addressing modern challenges. For example, an understanding of the tight regulation of senescence can be applied to modify the grain filling stage in an appropriate plant organism in order to increase the grain yield (harvest index). Altering the senescence stage to enhance remobilization or delay senescence through stay-green strategies (the most successful approaches being enhancing endogenous cytokine pathways and reducing ethylene production or perception) has become a routine approach to increasing productivity (Gan and Amasino 1995). Similarly senescence promoters and pathways have already been used to augment the flavour and deter the spoilage of products such as tomatoes and augment the nutrition of wheat.

Development of transcriptional data-sets such as the one we describe in tobacco will continue to facilitate discovery and drive innovation. Understanding how plants use nitrogen could potentially lead to improving nitrogen strategies that increase productivity of the plant and enhance the sustainability of farming. In the case of tobacco, knowledge of senescence and nitrogen metabolism is being applied to altering the leaf to decrease the level of target chemicals found in tobacco smoke. The extent to which plant genome initiatives are being undertaken by governments, academics and industrial partners will serve to ensure that genomics and the related branches of research will continue to contribute new tools, including genes and pathways that can regulate senescence and applications that promise to have an impact on modern society.

6. Acknowledgement

We would like to acknowledge Susie Davenport and Gwendoline Leach for the provision of data and Barbara Nasto for assistance with the writing/editing of this manuscript.

7. References

Alegre, H., & Munné-Bosch S. (2004). Drought-induced changes in flavonoids and other low molecular weight antioxidants in *Cistus clusii* grown under Mediterranean field conditions, *Tree Physiology*, 24(11) pp 1303-1311

Andersson, A., Keskitalo, J., Sjödin, A., Bhalerao, R., Sterky, F., Wissel, K., Tandre, K., Aspeborg, H., Moyle, R., Ohmiya, Y., Bhalerao, R., Brunner, A., Gustafsson, P., Karlsson, J., Lundeberg, J., Nilsson, O., Sandberg, G., Strauss, S., Sundberg, B., Uhlen, M., Jansson, S., & Nilsson, P. (2004) A transcriptional timetable of autumn senescence, *Genome Biology*, 5: R24

Baker, R. R. (1999) Smoke chemistry, In: *Tobacco Producton, Chemistry and Technology*, Davis D L & Nielsen M. T., pp. 398-439, Blackwell Science Ltd., ISBN 0-632-04791-7, Oxford

Bassham, D. C. (2007) Plant autophagy--more than a starvation response, *Current Opinion in Plant Biology*, 10(6) pp 587-593

Borgerding, M., & Klus, H. (2005) Analysis of complex mixtures--cigarette smoke. *Exp Toxicol Pathol*, 57(1) pp 43-73.

Bhalerao, R., Keskitalo, J., Sterky, F., Erlandsson, R., Björkbacka, H., Birve, S. J., Karlsson, J., Gardeström, P., Gustafsson, P., Lundeberg, J. & Jansson, S. (2003) Gene expression in autumn leaves. *Plant Physiology*, 131(2) pp 430-442.

British American Tobacco (n.d.) Available from www.BAT-science.com

Buchanan-Wollaston, V. & Ainsworth, C. (1997) Leaf senescence in *Brassica napus*: cloning of senescence related genes by subtractive hybridisation. *Plant Molecular Biology*, 33(5) pp 821-834

Buchanan-Wollaston, V., Page, T., Harrison, E., Breeze, E., Lim, P. O., Nam, H. G., Lin, J. F., Wu, S. H., Swidzinski, J., Ishizaki, K. & Leaver, C. J. (2005) Comparative transcriptome analysis reveals significant differences in gene expression and signalling pathways between developmental and dark/starvation-induced senescence in Arabidopsis, *Plant Journal*, 42 pp 567-585

Burton, H. R., Bush, L. P. & Djordjevic, M. V., (1989) Influence of temperature and humidity on the accumulation of tobacco-specific nitrosamines in stored burley tobacco. *Journal of Agriculture Food Chemistry* 37, pp 1372-1377

Burton, H. R., Dye, N. K. & Bush, L. P., (1994) Relationship between Tobacco-Specific Nitrosamines and Nitrite from Different Air-Cured Tobacco Varieties. *Journal of Agriculture Food Chemistry* 42 (9), pp 2007–2011

Causier, B., Kieffer, M., Davies, B. (2002) Plant biology. MADS-box genes reach maturity, *Science*, 296(5566) pp275-276.

Comai, L., Dietrich, R. A., Maslyar, D. J., Baden, F. S. & Harada, J. J., Coordinate expression of transcriptionally regulated isocitrate lyase and malate synthase genes in Brassica napus L., *Plant Cell*, 1989, 1(3) pp 293–300

Coque, M., Martin, A., Veyrieras, J. B., Hirel, B. & Gallais, A. (2008) Genetic variation for N-remobilization and postsilking N-uptake in a set of maize recombinant inbred lines. 3. QTL detection and coincidences, *Theoretical and Applied Genetics*, 117(5) pp 729-747

Chung, T., Suttangkakul, A. & Vierstra, R. D. (2009) The ATG autophagic conjugation system in maize: ATG transcripts and abundance of the ATG8-lipid adduct are regulated by development and nutrient availability, *Plant Physiology* 149, pp220-234

Day, A. & Goldschmidt-Clermont, M. (2011) The chloroplast transformation toolbox: selectable markers and marker removal *Plant Biotechnology Journal* 9 (5) pp 540-553

Dodd, A., Salathia, N., Hall, A., Kévei, E., Tóth, R., Nagy, F., Hibberd, J. M., Millar, A. J. & Webb, A. A. R. (2005), Plant Circadian Clocks Increase Photosynthesis, Growth, Survival, and Competitive Advantage. *Science* 309(5734) pp 630-633

Davis, L. D. & Neilsen, M (1999) Tobacco Production Chemistry Technology, Blackwell Science Ltd., ISBN 0-632-04791-7, Oxford

Edwards K. D., Bombarely, A., Story, G. W., Allen, F., Mueller, L. A., Coates, S. A., Jones, L. (2010) TobEA: an atlas of tobacco gene expression from seed to senescence, *BMC Genomics*, 11:142.

Edwards, K., Anderson, P., Hall, A., Salathia, N., Locke, J., Lynn, J., Straume, M., Smith, J. & Millar, A. (2006) FLOWERING LOCUS C Mediates Natural Variation in the High-Temperature Response of the Arabidopsis Circadian Clock, *Plant Cell*, 18 pp 639–650

Feys B.J., Wiermer, M., Bhat, R. A., Moisan, L. J., Medina-Escobar, N., Neu, C., Cabral, A. & Parker, J. E. (2005) Arabidopsis SENESCENCE-ASSOCIATED GENE101 stabilizes and signals within an ENHANCED DISEASE SUSCEPTIBILITY1 complex in plant innate immunity. *Plant Cell*, 17(9) pp 2601-2613 ischer, S., Spiegelhalder, B. & Preussmann, R. (1989) Preformed tobacco-specific nitrosamines in tobacco – role of nitrate and influence of tobacco type, *Carcinogenesis*, 10 pp1511-1517

racheboud, Y., Luquez, V., Björkén, L., Sjödin, A., Tuominen, H. & Jansson S. (2009) The control of autumn senescence in European aspen, *Plant Physiology*, 149(4) pp 1982-1891

riedrich, J. W. & Huffaker, R. C. (1980) Photosynthesis, leaf resistances, and ribulose-1,5-bisphosphate carboxylase degradation in senescing barley leaves, *Plant Physiology*, 65(6) pp 1103-1107

 an, S. & Amasino, R. M., (1995) Inhibition of Leaf Senescence by Autoregulated Production of Cytokinin, *Science* 270(5244) pp 1986-1988

hosh, S., Meli, V., Kumar, S., Thakur, A., Chakraborty, N., Chakraborty, S. & Datta, A. (2011) The N-glycan processing enzymes α-mannosidase and β-D-N-acetylhexosaminidase are involved in ripening-associated softening in the non-climacteric fruits of capsicum, *Journal of Experimetal Botany* 62(2) pp 571–582

rbic, V. & Bleecker, A. (1995) Ethylene regulates the timing of leaf senescence in Aradopsis. *The Plant Journal*, 8(4) pp 595-602

epstein, S., Sabehi, G., Carp, M. J. T. H., Falah, M., Nesher, O., Yariv, I., Dor, C. & Bassani, M. (2003) Large-scale identification of leaf senescence-associated genes. *The Plant Journal* 36 pp 629-642

regersen, P. L. & Holm, P. B. (2007) Transcriptome analysis of senescence in the flag leaf of wheat (*Triticum aestivum L.*), *Plant Biotechnology Journal* 5(1) pp 192-206.

regersen, P. L., Holm, P. B. & Krupinska, K. (2008) Leaf senescence and nutrient remobilisation in barley and Wheat, *Plant Biology*, 10 pp 37-49

riffith, C. M., Hosken, S. E., Oliver, D., Chojecki, J. & Thomas, H. (1997) Leaf senescence: Physiology and molecular biology, *Plant Molecular Biology*, 34 pp 815–821

uo, Y. & Gan, S. (2005) Leaf Senescence: Signals, Execution, and Regulation *Current Topics in Developmental Biology*. 71 pp 83-112

uo, Y., Cai, Z. & Gan, S. (2004) Transcriptome of Arabidopsis leaf senescence. *Plant Cell & Environment* 27 pp 521-549

armer, S. L., Hogenesch, J. B., Straume, M., Chang, H. S., Han, B., Zhu, T., Wang, X., Kreps, J. A. & Kay, S, A. (2000) Orchestrated transcription of key pathways in Arabidopsis by the circadian clock. *Science* 290(5499) pp 2110-2113

arrington, G. & Bush, D. (2003) The Bifunctional Role of Hexokinase in Metabolism and Glucose Signaling, *Plant Cell*, 15(11) pp 2493–2496

aussühl, K., Andersson, B. & Adamska, I. (2001) A chloroplast DegP2 protease performs the primary cleavage of the photodamaged D1 protein in plant photosystem II. *EMBO Journal* 20 pp 713–722

e, Y. & Gan, S. (2002) A gene encoding an acyl hydrolase is involved in leaf senescence in Arabidopsis, *Plant Cell*, 14(4) pp 805-815

e, Y., Tangm, W., Swain, J. D., Green, A. L., Jack, T. P. & Gan, S. (2001) Networking Senescence-Regulating Pathways by Using Arabidopsis Enhancer Trap Lines, *Plant Physiology*, 126 pp 707-716

ensel, L. L., Garbic, V., Baumgarten, D. A. & Bleecker, A. B., (1993) Leaf senescence: Physiology and molecular biology. *Plant Cell*, 5 pp 553–564.

Hoffmann, D. & Wynder, E. L. (1967) The reduction of the tumorigenicity of cigarette smoke condensate by addition of sodium nitrate to tobacco, *Cancer Research*, 27(1) pp 172-174

Hopkins, M., Taylor, C., Liu, Z., Ma, F., McNamara, L., Wang, T. W. & Thompson, J. E. (2007) Regulation and execution of molecular disassembly and catabolism during senescence. *New Phytologist*. 175 pp 201-214

Hörtensteiner, S. (2006) Chlorophyll degradation during senescence. *Annual Reviews in Plant Biology* 57 pp 55-77

Hörtensteiner, S. & Feller, U. (2002) Nitrogen metabolism and remobilization during senescence, *Journal of Experimental Botany* 53(370) pp 927-37

Jackson, S. D. (2009) Plant responses to photoperiod, *New Phytologist*, 181(3) pp 517-531

Klee, H. J. (2010) Improving Flavor of Fresh Fruits: Genomics biochemistry and biotechnology *New Phytologist*, 187(1) pp 44-56

Kim, D. J. & Smith, S. M. (1994) Molecular cloning of cucumber phosphoenolpyruvate carboxykinase and developmental regulation of gene expression, *Plant Molecular Biology*, 26 pp 423-434

Labboun, S., Tercé-Laforgue, T., Roscher, A., Bedu, M., Restivo, F. M., Velanis, C. N., Skopelitis, D. S., Moschou, P. N., Roubelakis-Angelakis, K. A., Suzuki, A. & Hirel, B. (2009) Resolving the role of plant glutamate dehydrogenase. I. In vivo real time nuclear magnetic resonance spectroscopy experiments. *Plant Cell Physiology*, 50(10) pp 1761-1773

Lea, P. J. & Miflin, B. J., (1974) Alternative route for nitrogen assimilation in higher plants. *Nature*. 251(5476) pp 614-616

Lim, P. O., Kim, H. J. & Nam, H. G. (2007) Leaf senescence. *Annual Review of Plant Biology*. 58 pp 115-136

Lin, J. F. & Wu, S. H. (2004) Molecular events in senescing Arabidopsis leaves, *Plant Journal*, 39(4) pp612-628

Lohman, K., Gan, S., John, M. & Amasino, R. M. (1994) Molecular analysis of natural leaf senescence in *Arabidopsis thaliana*, *Physiologia Plantarum*, 92 pp 322-328

Lossl, A. & Waheed, M. (2011) Chloroplast-derived vaccines against human diseases: achievements, challenges and scopes, *Plant Biotechnology Journal*, 9 pp 527-539

Mae, T., Kai, N., Makino, A. & Ohira, K. (1984) Relation between ribulose bisphosphate carboxylase content and chloroplast number in naturally senescing primary leaves of wheat. *Plant Cell & Physiology*, 25 pp 333-336

Masclaux, C., Valadier, M. H., Brugière, N., Morot-Gaudry, J. F. & Hirel, B. (2000) Characterization of the sink/source transition in tobacco (Nicotiana tabacum L.) shoots in relation to nitrogen management and leaf senescence, *Planta*, 211(4) pp 510-518

Morris, K., MacKerness, S. A., Page, T., John, C. F., Murphy, A. M., Carr, J. P. & Buchanan-Wollaston, V. (2000) Salicylic acid has a role in regulating gene expression during leaf senescence, *Plant Journal*, 23(5) pp 677-685

Mueller, L. A., Solow, T. H., Taylor, N., Skwarecki, B., Buels, R., Binns, J., Lin, C., Wright, M. H., Ahrens, R., Wang, Y., Herbst, E. V., Keyder, E. R., Menda, N., Zamir, D. & Tanksley, S. D. (2005) The SOL Genomics Network. A Comparative Resource for Solanaceae Biology and Beyond. *Plant Physiol.* 138(3) pp 1310-1317.

Noodén, L. D., Guiamét, J. J. & John, I. (1997) Senescence mechanisms, *Physiologia Plantarum*, 101 pp 746-753

Otegui, M. S., Noh, Y. S., Martínez, D. E., Vila Petroff, M. G., Staehelin, L. A., Amasino & R. M., Guiamet, J. J. (2005) Senescence-associated vacuoles with intense proteolytic activity develop in leaves of Arabidopsis and soybean. *Plant Journal*, 41(6) pp 831-844

Norman, A. (1999). Cigarette Design and Materials. In: *Tobacco Producton, Chemistry and Technology*, Davis D L & Nielsen M. T., pp. 353-387, Blackwell Science Ltd., ISBN 0-632-04791-7, Oxford

Ohsumi, Y. (2001) Molecular mechanism of bulk protein degradation in lysosome/vacuole, *Tanpakushitsu Kakusan Koso*, 46(11) pp 1710-1716

Pruzinska, A., Tanner, G., Aubry, S., Anders, I., Moser, S., Muller, T., Ongania, K. H., Krautler, B., Youn, J. Y., Liljegren, S. J. & Hortensteiner, S. (2005) Chlorophyll Breakdown in Senescent Arabidopsis Leaves. Characterization of Chlorophyll Catabolites and of Chlorophyll Catabolic Enzymes Involved in the Degreening Reaction. *Plant Physiology*, 139 pp 52-63

Quirino, B. F., Normanly, J. & Amasino, R. M. (1999) Diverse range of gene activity during Arabidopsis thaliana leaf senescence includes pathogen-independent induction of defense-related genes. *Plant Molecular Biology*, 40(2) pp 267-278.

Rodgman, A. & Perfetti, T. A. (2009) The chemical components of tobacco and tobacco smoke. CRC press, Florida, USA. pp 1259. ISBN 978-1-4200-7883-1.

Shaul, O. & Galili, G. (1993) Concerted regulation of lysine and threonine synthesis in tobacco plants expressing bacterial feedback-insensitive aspartate kinase and dihydrodipicolinate synthase. *Plant Molecular Biology*, 23(4) pp 759-768.

Smart, C. M. (1994) Gene expression during leaf senescence. *New Phytologist* 126 pp 419-448

Sokolenko, A., Lerbs-Mache, S., Altschmied, L., Herrmann, R. G. (1998) Clp protease complexes and their diversity in chloroplasts. *Planta*, 207 pp 286-295

Spiegelhalder, B., Bartsch, H.(1996) Tobacco-specific nitrosamines, *European Journal of Cancer Prevention*, 5 pp 33-38

Stitt, M., Müller, C., Matt, P., Gibon, Y., Carillo, P., Morcuende, R., Scheible, W. R. & Krapp, A. (2002) Steps towards an integrated view of nitrogen metabolism, *Journal of Experimental Botany* 53(370) pp 959-70

Thomas, H. & Donnison, I. (2000) Back from the brink: plant senescence and its reversibility, *Symposium Society of Experimental Biology*, 52 pp 149-162

Thomas, H. & Stoddart, J. L. (1980). Leaf senescence, *Annual Review of Plant Physiology*, 31 pp 83-111

Thompson, A. R. & Vierstra, R. D. (2005) Autophagic recycling: lessons from yeast help define the process in plants, *Current Opinions in Plant Biology*, 8(2) pp 165-173

Tollenaar, M. & Wu, J. (1999). Yield improvement in temperate maize is attributable to greater stress Tolerance, *Crop Science*, 39 pp 1597-1604

Uauy, C., Distelfeld, A., Fahima, T., Blechl, A. & Dubcovsky J (2006) A NAC Gene regulating senescence improves grain protein, zinc, and iron content in wheat *Science*, 314(5803) pp 1298-1301

Van der Graaff, E., Schwacke, R., Schneider, A., Desimone, M., Flugge, U. I. & Kunze, R. (2006) Transcription Analysis of Arabidopsis Membrane Transporters and Hormone Pathways during Developmental and Induced Leaf Senescence, *Plant Physiology*, 141 pp 776-792

Van Doorn, W. G. (2008) Is the onset of senescence in leaf cells of intact plants due to low or high sugar levels? *Experimental Botany* 59(8) pp 1963-1972

Vierstra, R. D. (1996) Proteolysis in plants: mechanisms and functions. *Plant Molecular Biology*, 32(1-2) pp 275-302

Verburg, J. G., Huynh, Q.K. (1991) Purification and Characterization of an Antifungal Chitinase from Arabidopsis thaliana, *Plant Physiology*, 95(2) pp 450-455

Watanabe, A., Hamada, K., Yokoi, H. & Watanabe, A., (1994) Biophysical and differential expression of cytosolic glutamine synthetase genes of radish during seed germination and senescence of cotyledons, *Plant Molecular Biology*, 26 pp 1807–1817

Weasel, L. H. (2009) *Food Fray: Inside the Controversy over Genetically Modified Food*, Amacom, ISBN-13: 978-0814401644, New York

Wingler, A. & Roitsch, T. (2008) Metabolic regulation of leaf senescence: interactions of sugar signalling with biotic and abiotic stress responses, *Plant Biology*, 10 pp 50-62.

Wittenbach, V. A. (1978) Breakdown of Ribulose Bisphosphate Carboxylase and Change in Proteolytic Activity during Dark-induced Senescence of Wheat Seedlings, *Plant Physiology*, 62(4) pp 604-608

Wu, W., Zhang, L., Jain, R. B., Ashley, D. L. & Watson, C. H. (2005) Determination of carcinogenic tobacco-specific nitrosamines in mainstream smoke from U.S.-brand and non-U.S.-brand cigarettes from 14 countries, *Nicotine Tobacco Research*, 7(3) pp 443-451

Xue-Xuan, X., Hong-Bo, S., Yuan-Yuan, M., Gang, X., Jun-Na, S., Dong-Gang, G. & Cheng-Jiang, R. (2010) Biotechnological implications from abscisic acid (ABA) roles in cold stress and leaf senescence as an important signal for improving plant sustainable survival under abiotic-stressed conditions, *Critical Reviews in Biotechnology*, 30(3) pp 222-230

Yang, J. & Zhang, J. (2006) Grain filling of cereals under soil drying, *New Phytologist* 169(2) pp 223-236

Yoshida, S. (2003) Molecular regulation of leaf senescence, *Current opinion in plant biology*, 6 pp 79-84

Zavaleta-Mancera, H. A., Thomas, B. J., Thomas, H. & Scott, I. M. (1999) Regreening of senescent Nicotiana leaves, *Journal of Experimental Botany*, 50(340) pp 1683–1689

3

Photo- and Free Radical-Mediated Oxidation of Lipid Components During the Senescence of Phototrophic Organisms

Jean-François Rontani
Laboratory of Microbiology, Geochemistry and Marine Ecology (UMR 6117),
Center of Oceanology of Marseille, Aix-Marseille University, Campus of Luminy, Marseille,
France

1. Introduction

Recently, the role played by photochemical and free radical-mediated processes in the degradation of lipid components during the senescence of phototrophic organisms was investigated. The present paper reviews the results obtained in the course of these studies.

In a first part, visible and UV light-induced photooxidation of the main lipid cell components (chlorophylls, carotenoids, sterols, unsaturated fatty acids, highly branched isoprenoid and linear alkenes, alkenones, cuticular waxes ...) in senescent phototrophic organisms (phytoplankton, cyanobacteria, higher plants, purple sulfur bacteria and aerobic anoxygenic phototrophic bacteria) is examined. Probably due to its long lifetime in hydrophobic micro-environments and thus in senescent cells, singlet oxygen plays a key role in the photodegradation of most of the lipid components.

The second part of this paper describes the free radical oxidation (autoxidation) of lipid components during the senescence of phototrophic organisms, which have been virtually ignored until now in the literature. In senescent phototrophic organisms, the mechanism of initiation of free-radical oxidation seems to be the homolytic cleavage (catalyzed by some metal ions) of photochemically produced hydroperoxides. It was also demonstrated recently that viral infection and autocatalytic programmed cell death could also lead to elevated production of reactive oxygen species (ROS) able to induce the degradation of cell components.

2. Photodegradation processes in phototrophic organisms

Several works suggested photo-oxidation as an important sink of organic matter in the photic layer of oceans (Zafiriou, 1977; Zafiriou et al., 1984). However, due to the lack of suitable markers this phenomenon has never been fully addressed. Owing to the problem of stratospheric ozone depletion, some studies have recently examined the degradative effects of enhanced UV-B doses on phytoplanktonic lipids (He and Häder, 2002). However, photochemical damages in phytoplanktonic cells are not a monopoly of UV radiation. In fact, due to the presence of chlorophylls (which are very efficient photosensitizers (Foote, 1976; Knox and Dodge, 1985)), numerous organic components of phytoplankton are susceptible to being photodegraded during senescence by photosynthetically active radiation (PAR).

2.1 Photodegradation of the main lipidic components of phytoplankton during senescence

When a chlorophyll molecule absorbs a quantum of light energy, an excited singlet state (^{1}Chl) is formed which, in healthy cells, leads predominantly to the characteristic fast reactions of photosynthesis (Foote, 1976). However, a small proportion (<0.1%) undergoes intersystem crossing (ISC) to form the longer lived triplet state (^{3}Chl; Knox and Dodge, 1985). ^{3}Chl is not only itself potentially damaging in type I reactions (hydrogen atom or electron abstraction) (Knox and Dodge, 1985), but can also generate highly reactive oxygen species (ROS) and, in particular, singlet oxygen (^{1}O$_2$), by reaction with ground state oxygen (^{3}O$_2$) via Type II processes. In order to avoid oxidative damage, there are many antioxidant protective mechanisms in chloroplasts. Carotenoids quench ^{3}Chl and ^{1}O$_2$ by energy transfer mechanisms at very high rates (Foote, 1976) and tocopherols can remove ^{1}O$_2$, O$_2$$^{•-}$, HOO$^•$ and HO$^•$ by acting as sacrificial scavengers (Halliwell, 1987). Superoxide dismutase enzyme (SOD) and ascorbic acid may also scavenge O$_2$$^{•-}$ (Halliwell, 1987), while catalase activity decreases H$_2$O$_2$ levels.

In senescent phototrophic organisms, the fast reactions of photosynthesis clearly do not operate, so an accelerated rate of formation of ^{3}Chl and ^{1}O$_2$ would be expected (Nelson, 1993). The rate of formation of these potentially damaging species can then exceed the quenching capacity of the photoprotective system and photodegradation can occur (photodynamic effect; Merzlyak and Hendry, 1994). In phytodetritus, when the ordered structure of the thylakoid membranes has been disrupted, pigments tend to remain associated with other hydrophobic cellular components such as membrane lipids (Nelson, 1993). As a result, the photooxidative effect of chlorophyll sensitization might be strongly amplified within such a hydrophobic micro-environment. Moreover, the lifetime of ^{1}O$_2$ produced from sensitizers in a lipid-rich hydrophobic environment could be longer, and its potential diffusive distance greater, than its behaviour in aqueous solution (Suwa et al., 1977). It is not surprising, therefore, that photodegradation processes act on the majority of unsaturated lipid components of senescent phytoplankton.

2.1.1 Chlorophylls

Irradiation of dead phytoplankton cells by PAR and UVR radiations results in rapid degradation of chlorophylls (Nelson, 1993; Rontani et al., 1995; Christodoulou et al., 2010). Photodegradation of chlorophyll-a and -c in killed cells of E. huxleyi appeared to be induced by both PAR and UVR (Christodoulou et al., 2010). The photochemical degradation of chlorophylls has so far been studied almost exclusively with respect to the macrocycle moiety of the molecule, which is the more reactive. Despite some progress regarding intermediary photoproducts (Engel et al., 1991; Iturraspe et al., 1994), no stable and specific markers for the chlorophyll macrocycle photodegradation have been characterised.

The isoprenoid phytyl side-chain of chlorophylls is also sensitive to photochemical processes. In fact, in phytodetritus, the photodegradation rates were only 3 to 5 times higher for the chlorophyll tetrapyrrolic structure than for the phytyl side-chain (Cuny et al., 1999; Christodoulou et al., 2010). Analysis of isoprenoid photoproducts of chlorophylls after irradiation of different dead phytoplanktonic cells by visible light clearly established that the photodegradation of the chlorophyll phytyl side-chain in phytodetritus involved mainly

$^1O_2$. The type II (i.e. involving $^1O_2$) photosensitized oxidation of the phytol moiety of chlorophylls leads to the production of photoproducts of structures **a** and **b** (Fig. 1), quantifiable after NaBH₄-reduction and alkaline hydrolysis respectively in the form of 6,10,14-trimethylpentadecan-2-one (**1**) (phytone) and 3-methylidene-7,11,15-trimethyl-hexadecan-1,2-diol (phytyldiol) (**2**) (Fig. 1) (Rontani et al., 1994).

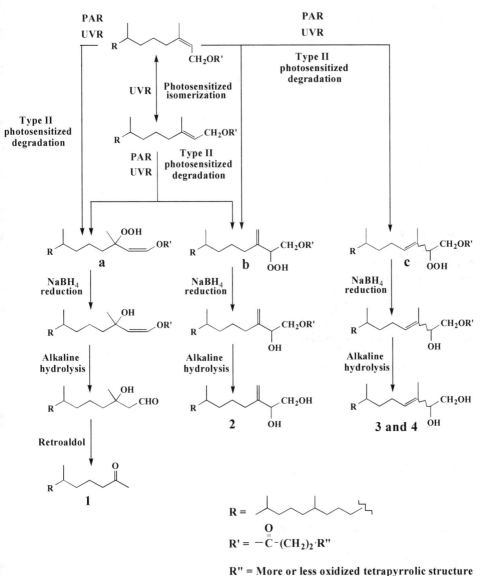

Fig. 1. Photooxidation of chlorophyll phytyl side-chain and reactions of oxidation products during alkaline hydrolysis.

Irradiation with UVR resulted in the additional production of small amounts of Z-phytol and Z and E-3,7,11,15-tetramethylhexadec-3-en-1,2-diols (**3,4**) (Christodoulou et al., 2010). The detection of Z-phytol allowed to demonstrate the induction of *cis-trans* photosensitized isomerization by UVR. These reactions probably involve triplet states of ketones as sensitizers. Type II photosensitized oxidation of the Z configuration of phytol, which should lead to the production of photoproducts of structures **a**, **b** and **c** (Fig. 1) (Schulte-Elte et al., 1979), explains the detection of small amounts of Z and E-3,7,11,15-tetramethylhexadec-3-en-1,2-diols (**3,4**) after irradiation with UVR. Irradiation with UVR also resulted in a faster degradation of chlorophyll phytyl side-chain oxidation products (Christodoulou et al., 2010). This higher reactivity was attributed to UVR-induced homolysis of the peroxyl group of photoproducts of structures **a**, **b** and **c** (Fig. 1).

Phytyldiol (**2**) is ubiquitous in the marine environment and has been proposed as tracer for photodegradation of chlorophyll's phytyl side chain (Rontani et al. 1994; 1996a; Cuny and Rontani 1999). Further, the molar ratio phytyldiol:phytol (Chlorophyll Phytyl side-chain Photodegradation Index, CPPI) was employed to estimate the extent of chlorophyll photodegraded in natural marine samples by the empirical equation: chlorophyll photodegradation % = $(1-(CPPI + 1)^{-18.5}) \times 100$ (Cuny et al. 2002).

2.1.2 Carotenoids

In phytodetritus, chlorophylls and carotenoids remain in a close molecular-scale association at relatively high localized concentrations, even though the structure of the thylakoid membrane has been disrupted (Nelson, 1993). Thus, the sensitized photooxidation of carotenoids is enhanced. The photosensitized oxidation (involving $^1O_2$) of carotenoids in solvents has been studied (Iseo et al., 1972) and loliolide (**5**), *iso*-loliolide (**6**) and dihydroactinidiolide (**7**) (Fig. 2) were identified as major photoproducts, depending on the functionality of carotenoids at C-3. Loliolide (**5**) and *iso*-loliolide (**6**) have been detected in killed cells of *Dunaliella* sp. irradiated by visible light (Rontani et al., 1998). However, due to their apparent production by anaerobic bacteria (Repeta, 1989) and during dark incubations of killed phytoplanktonic cells (Rontani et al., 1998), these compounds cannot constitute unequivocal indicators of photooxidative processes.

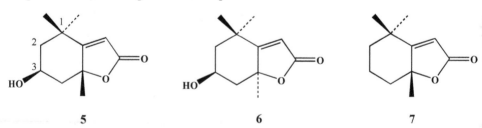

| 5 | 6 | 7 |

Fig. 2. Structure of the main carotenoid oxidation products.

2.1.3 Δ⁵-sterols

As important unsaturated components of biological membranes, Δ⁵-sterols are highly susceptible to photooxidative degradation during the senescence of phytoplankton. Irradiation by visible light of killed cells of *Skeletonema costatum*, *Dunaliella* sp.,

Phaeodactylum tricornutum and *Emiliania huxleyi* (Rontani et al., 1997a; 1997b; 1998) resulted in a quick photodegradation of the sterol components of these algae. The results obtained clearly established that the photooxidation of sterols in senescent cells of phytoplankton involves type II photoprocesses. These processes mainly produce Δ^6-5α-hydroperoxides (**8**) and to a lesser extent Δ^4-6α/6β-hydroperoxides (**9** and **10**) (Fig. 3) (Nickon and Bagli, 1961; Kulig and Smith, 1973). Δ^6-5α-hydroperoxysterols (**8**) are relatively unstable and may undergo allylic rearrangement to Δ^5-7α-hydroperoxysterols (**11**), which in turn epimerize to the corresponding 7β-hydroperoxides (**12**) (Fig. 3) (Smith, 1981). It was previously demonstrated that during singlet oxygen-mediated photooxidation of sterols in biological membranes (Korytowski *et al.*, 1992) and senescent phytoplanktonic cells (Rontani *et al.*, 1997a) the photogeneration of Δ^4-6α/6β-hydroperoxides (**9** and **10**) was more favourable than in homogeneous solution (ratio Δ^4-6α/6β-hydroperoxides/Δ^6-5α-hydroperoxysterols ranging from 0.30 to 0.35 instead of 0.1).

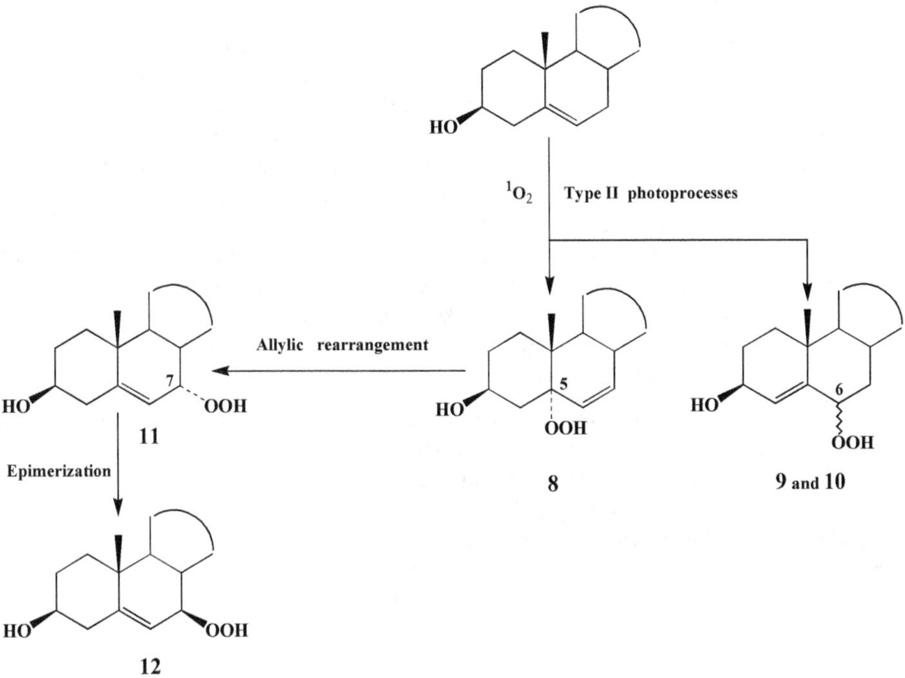

Fig. 3. Type II photosensitized oxidation of Δ^5 sterols.

Allylic rearrangement of Δ^6-5α-hydroperoxides (**8**) appeared to take place very weakly in senescent phytoplanktonic cells (Rontani et al., 1997a; 1997b; 1998). This surprising stability was attributed by Korytowski et al. (1992) either to hydrogen bonding between the unsaturated fatty acyl chain of phospholipids and Δ^6-5α-hydroperoxides (**8**) which could hinder the allylic rearrangement, or to differences of polarity in the carbon 7-10 zone of the fatty acyl chain (where sterols tend to localize in phospholipid/sterol bilayers (MacIntosch, 1978)). It is also interesting to note that the reduction of hydroperoxysterols to the corresponding diols weakly operates in killed phytoplanktonic cells (Rontani et al., 1997a).

Δ^6-5α-Hydroperoxysterols (8) are potential type II photodegradation markers, not only because they are the major products of singlet oxygen attack on the steroidal Δ^5-3β- system, but also because biological functionalization of steroids at C-5 is rare. Unfortunately, if these compounds are particularly stable in phytodetritus, they decay slowly in the sediment to their corresponding Δ^5-7α/β-derivatives (11 and 12) (Rontani and Marchand, 2000), which are not selective markers (see chapter 3.3). Moreover, according to the stability of the alkyl radicals formed during β-scission of the corresponding alkoxyl radicals, the following order of stability was proposed: Δ^4-6-hydroperoxysterols (9 and 10) > Δ^5-7-hydroperoxysterols (11 and 12) > Δ^6-5-hydroperoxy-sterols (8) (Christodoulou et al., 2009). Consequently, Δ^4-6α/β-hydroperoxysterols (9 and 10) (or their degradative products Δ^4-6α/β-hydroxysterols and Δ^4-6α/β-oxosterols) may be considered as more reliable *in situ* markers of type II photodegradation processes than Δ^6-5α-hydroperoxides (8).

2.1.4 Unsaturated fatty acids

Chloroplast membrane components are particularly susceptible to type II photooxidation (Heath and Packer, 1968). This is the case notably for unsaturated fatty acids, which generally predominate in algal lipids, particularly in the photosynthetic membranes (Woods, 1974). In killed phytoplanktonic cells, the photodegradation rates of unsaturated fatty acids logically increase with their unsaturation degree (Rontani et al., 1998). Singlet oxygen-mediated photooxidation of monounsaturated fatty acids involves a direct reaction of $^1O_2$ with the carbon–carbon double bond by a concerted 'ene' addition (Frimer 1979) and leads to formation of hydroperoxides at each carbon of the original double bond. Thus, photooxidation of oleic acid produces a mixture of 9- and 10-hydroperoxides with an allylic *trans*-double bond (Frankel et al. 1979; Frankel, 1998), which can subsequently undergo stereoselective radical allylic rearrangement to 11-*trans* and 8-*trans* hydroperoxides, respectively (Porter et al. 1995) (Fig. 4).

The free radical nature of the allylic hydroperoxide rearrangement is supported by the observation that the rearrangement is catalysed by free radical initiators or light and inhibited by phenolic antioxidants (Porter et al., 1995). This allylic rearrangement weakly intervenes in most of the killed phytoplanktonic cells examined (Rontani et al., 1998). This was attributed to the relatively high localized fatty acid concentrations present in phytodetritus (Nelson, 1993), which favoured the dimerisation of hydroperoxides. Hydrogen atom abstraction to form allylperoxyl radicals does indeed occur readily from hydroperoxide monomers but not from hydroperoxide dimers (Porter et al., 1995).

During early diagenesis, isomeric hydroperoxyacids undergo heterolytic cleavage to aldehydes and ω-oxocarboxylic acids (Frimer, 1979) or homolytic cleavage and subsequent transformation to the corresponding alcohols or ketones (Fig. 5).

Taking into account the high amounts of photoproducts of mono-unsaturated fatty acids detected in the particulate matter samples (Marchand and Rontani, 2001; Christodoulou et al., 2009; Rontani et al., 2011a), and the well known increasing photooxidation rates of fatty acids with their degree of unsaturation (Frankel., 1998), it can be concluded that considerable amounts of poly-unsaturated fatty acids must be photooxidized during the senescence of phytoplankton in the marine environment. However, at this time photooxidation products of this kind of fatty acids could not be detected in natural samples.

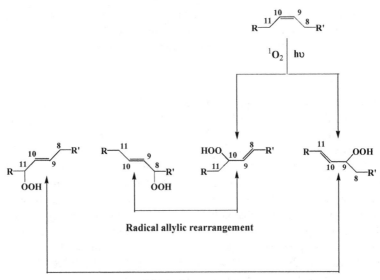

R = -(CH$_2$)$_6$-CH$_3$

R' = -(CH$_2$)$_6$-COOH

Fig. 4. Type II photosensitized oxidation of oleic acid.

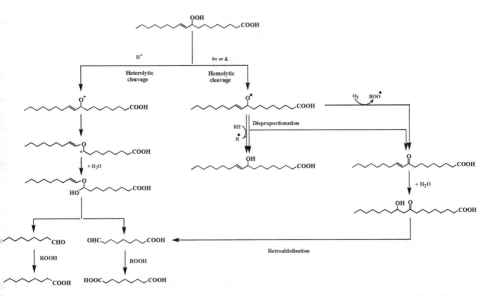

Fig. 5. Degradation of allylic hydroperoxides resulting from Type II photosensitized oxidation of monounsaturated fatty acids (the example given is this of 9-hydroperoxyoctadec-10-enoic acid) (RH = hydrogen donors, e.g. lipids or reduced sensitizers).

This is possibly due to: (i) the instability of the hydroperoxides formed, or (ii) the involvement of cross-linking reactions leading to the formation of macromolecular structures (Neff et al., 1988) non-amenable by gas chromatography.

2.1.5 Alkenones

Alkenones are a class of mono-, di-, tri-, tetra- and penta-unsaturated C_{35}-C_{40} methyl and ethyl ketones (Boon et al., 1978; Volkman et al., 1980; de Leeuw et al., 1980; Marlowe et al., 1984; Prahl et al., 2006; Jaraula et al., 2010), which are produced by certain marine haptophytes. *Emiliania huxleyi* and *Gephyrocapsa oceanica* are the major sources of alkenones in the open ocean (Volkman et al., 1980; 1995; Conte et al., 1994). The unsaturation ratio of C_{37} alkenones, defined as $U^{K'}_{37} = [C_{37:2}]$ / $([C_{37:2}] + [C_{37:3}])$ where $[C_{37:2}]$ and $[C_{37:3}]$ are the concentrations of di- and tri-unsaturated C_{37} alkenones respectively, varies positively with the growth temperature of the alga (Prahl and Wakeham, 1987; Prahl et al., 1988). The $U^{K'}_{37}$ - growth temperature relationship in haptophyte algae and transferred to sinking marine particulate matter leads to a linear relationship between sedimentary C_{37} alkenone composition and mean annual SST records throughout the oceans (Rosell-Melé et al., 1995; Müller et al., 1998). The $U^{K'}_{37}$ index is now routinely used for paleotemperature reconstruction.

For alkenones to be useful as measures of sea surface temperature in the geological record, it is essential that any effects of degradation in the water column and in sediments either do not affect the temperature signal established during their initial biosynthesis by the alga (Harvey, 2000; Grimalt et al., 2000), or if there is a change its extent can be reasonably estimated.

Visible light-induced photodegradation of these compounds was thus previously investigated in order to determine if photochemical processes could appreciably modify $U^{K'}_{37}$ ratios during algal senescence (Rontani et al., 1997b; Mouzdahir et al., 2001; Christodoulou et al., 2010). Though potentially selective, photochemical degradation of alkenones is not fast enough in killed cells of E. *huxleyi* to induce strong modifications of the $U^{K'}_{37}$ ratio before the photodestruction of the photosensitizing substances (Rontani et al., 1997b; Mouzdahir et al., 2001). UVR also appeared to be inefficient to alter the $U^{K'}_{37}$ ratio (Christodoulou et al., 2010).

This stability was attributed to the *trans* configuration of alkenone double bonds (Rechka and Maxwell, 1988) that is 7 to 10 times less sensitive against singlet oxygen-mediated oxidation than the classical *cis* configuration of fatty acids (Hurst et al., 1985). This may explain the difference of photoreactivity observed between the alkenones and fatty acids with the same number of unsaturations. We also previously attributed the poor photoreactivity of alkenones to a localisation of these compounds elsewhere than in cell membranes (Rontani et al., 1997b; Mouzdahir et al., 2001), which could significantly decrease the likelihood of interaction between singlet oxygen and alkenones. Although this hypothesis is well supported by the recent results of Eltgroth et al. (2005), who demonstrated that alkenones are mainly localized into cytoplasmic vesicles, the migration of singlet oxygen from phytodetritus to attached heterotrophic bacteria previously observed (Rontani et al., 2003a; Christodoulou et al., 2010) strongly suggests a diffusion of this excited form of oxygen also in these cytoplasmic vesicles.

2.1.6 n-Alkenes

The visible light-induced degradation of n-alkenes was previously investigated in killed cells of the Prymnesiophycea E. *huxleyi* and the Eustigmatophycea *Nannochloropsis salina* (Mouzdahir et al., 2001).

In E. *huxleyi* killed cells, minor C_{31} and C_{33} n-alkenes were strongly photodegraded, while the major C_{37} and C_{38} n-alkenes appeared particularly recalcitrant towards photochemical processes. These strong differences of photoreactivity imply distinct biological syntheses and/or functions for these two groups of hydrocarbons in E. *huxleyi* cells. Interestingly, the stereochemistry of the internal double bonds in C_{31} and C_{33} n-alkenes has been established to be *cis*, while C_{37} and C_{38} alkenes internal double bonds exhibit a *trans* geometry (Rieley et al., 1998; Grossi et al., 2000). The photochemical recalcitrance of C_{37} and C_{38} n-alkenes could thus be partly attributed to the *trans* geometry of their internal double bonds.

Irradiation of dead cells of N. *salina* resulted in a strong modification of the hydrocarbon fraction. It did not provide evidence of a significant light-dependent degradation of monounsaturated hydrocarbons; this result was attributed to the terminal position of the double bond in these compounds (Gelin et al., 1997), which is poorly reactive towards singlet oxygen (Hurst et al., 1985). In contrast, di-, tri-, and tetraenes were strongly photodegraded during irradiation. The visible light-dependent degradation of phytoplanktonic n-alkenes showed apparent second-order kinetics with respect to light exposure and the half-life doses obtained logically decrease with increasing number of double bonds in these compounds (Mouzdahir et al., 2001).

2.1.7 Highly branched isoprenoid (HBI) alkenes

HBI alkenes are widely distributed in aquatic environments (Rowland and Robson, 1990; Sinninghe-Damsté et al., 2004), although they appear to originate from a relatively small number of diatomaceous algae including *Haslea* spp., *Rhizosolenia* spp., *Pleurosigma* spp. and *Navicula* spp. (Volkman et al., 1994; Sinninghe-Damsté et al., 2004; Belt et al., 2000, 2001; Allard et al., 2001; Grossi et al., 2004). Despite this, they have been commonly reported in marine sediments worldwide and provide some insight into the deposition of organic matter from the water column. One HBI alkene, a mono-unsaturated isomer termed IP_{25}, has been used as a proxy for the occurrence of spring sea ice in the Arctic (e.g. Belt et al., 2007, 2010; Massé et al., 2008).

Examination of the photoreactivity of several mono-, di-, tri- and tetra-unsaturated HBI alkenes in the presence of a photosensitizer solution and in dead cells of H. *ostrearia* allowed to show that HBI alkenes possessing at least one tri-substituted double bond may be photo-oxidized at similar or higher rates compared to other highly reactive lipids (e.g. PUFAs, vitamin E and chlorophyll a) during the senescence of diatom cells (Rontani et al., 2011b). As a consequence, it is proposed that HBI alkenes possessing trisubstituted double bonds are likely to be susceptible to photodegradation within the euphotic zone. In contrast, HBIs containing only mono- and di-substituted double bonds were found to be significantly less reactive towards $^1O_2$ and should, therefore, be relatively preserved during sedimentation through the water column (Rontani et al., 2011b). The kinetic experiments are supported by product analysis, which revealed that the main reaction with $^1O_2$ primarily occurs with the trisubstituted double bonds of HBI alkenes affording tertiary and secondary allylic hydroperoxides (Fig. 6). In contrast, the extremely low photoreactivity of the HBI monoene

IP$_{25}$, can be attributed to its containing only the least photochemically reactive double bond. This lack of reactivity supports (in part) the good preservation of IP$_{25}$ generally observed in sediments (Belt et al., 2007, 2010; Massé et al., 2008).

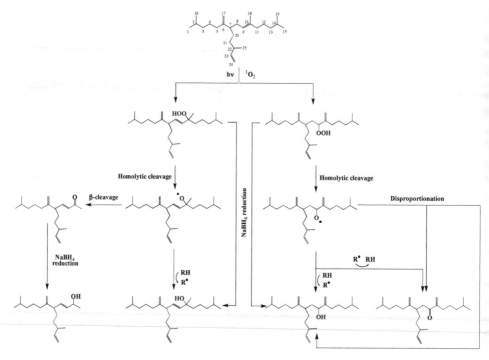

Fig. 6. Type II photosensitized oxidation of HBI alkenes (RH = hydrogen donors)

2.2 Photodegradation processes in other phototrophic organisms

Visible light-dependent degradation processes have been also studied in senescent cells of two purple sulfur bacteria (*Thiohalocapsa halophila* and *Halochromatium salexigens*) isolated from microbial mats from Camargue (France) (Marchand and Rontani, 2003). These reactions act intensively on the phytyl side chain of bacteriochlorophyll-*a* and lead to the production of phytone (**1**) and phytyldiol (**2**) as in the case of chlorophylls (Fig. 1). Palmitoleic and *cis*-vaccenic acids also undergo strong photodegradation, affording mainly isomeric allylic oxo-, hydroxy- and hydroperoxyacids.

These processes were also investigated in aerobic anoxygenic phototrophic bacteria (AAPs) (Rontani et al., 2003a). These organisms constitute a relatively recently discovered bacterial group (Yurkov and Beatty, 1998) and seem to be widespread in the open ocean (Kolber et al., 2000). They perform photoheterotrophic metabolism, requiring organic carbon for growth, but they are capable to use photosynthesis as an auxiliary source of energy (Kolber et al., 2001). Though sensitive to photochemical processes in senescent purple sulfur bacteria (Marchand and Rontani., 2003), the isoprenoid phytyl side-chain of bacteriochlorophyll -*a* is not significantly photodegraded in senescent cells of AAPs (Rontani et al., 2003a). In contrast, significant amounts of allylic hydroxyacids arising from the photo-oxidation of the major unsaturated fatty acid of these organisms (cis-vaccenic acid) could be detected after irradiation (Rontani et al., 2003a).

As in the case of phytoplankton and cyanobacteria, visible light-dependent degradation processes act significantly on the chlorophyll phytyl side-chain (Rontani et al., 1996b), unsaturated fatty acids and sterols (Rontani, Unpublished results) during terrestrial higher plant senescence affording similar photoproducts. 9-Hydroperoxy-18-hydroxyoctadec-10(trans)-enoic (13) and 10-hydroperoxy-18-hydroxyoctadec-8(trans)-enoic (14) acids deriving from type II photooxidation of 18-hydroxyoleic acid (15) (Fig. 7) were detected after visible light-induced senescence experiments carried out with Petroselinum sativum and subsequent cutin depolymerisation (Rontani et al., 2005a).

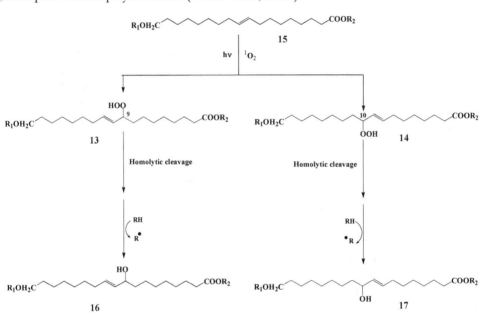

Fig. 7. Type II photosensitized oxidation of 18-hydroxyoleic acid in cutin polymers.

These results showed that in senescent plants, where the $^1O_2$ formation rate exceeds the quenching capacity of the photoprotective system, $^1O_2$ can migrate outside the chloroplasts and affect the unsaturated components of cutins. Significant amounts of 9,18-dihydroxyoctadec-10(trans)-enoic (16) and 10,18-dihydroxyoctadec-8(trans)-enoic (17) acids resulting from the reduction of these photoproducts of 18-hydroxyoleic acid were also detected in different natural samples (Rontani et al., 2005a). These results well support the significance of the photooxidation of the unsaturated components of higher plant cutins in the natural environment.

3. Free radical degradation (autoxidation) processes in phototrophic organisms

Autoxidation is the direct reaction of molecular oxygen with organic compounds under mild conditions. The autoxidation of organic compounds (in particular, lipids) involves free radical reaction chains and thus includes an initiation, a propagation and a termination phase. Mechanisms of initiation for the free radical processes have been the subject of many studies. In senescent phytoplanktonic cells, initiation seems to result from the decomposition of hydroperoxides produced during photodegradation of cellular organic matter (Rontani et al., 2003b). Until now, autoxidative degradation in the marine environment has been largely ignored. Specific markers of these reactions have been highlighted by *in vitro* studies (Frankel, 1998; Rontani et al., 2003b; Rontani and Aubert, 2005). Using these markers, it was demonstrated *in situ* that autoxidation plays a very significant role in the degradation of particulate organic matter (Marchand et al., 2005; Rontani et al., 2006; Christodoulou et al., 2009; Rontani et al., 2011a).

Although the occurrence of autoxidation processes was clearly demonstrated *in situ*, it is not easy to induce these processes in laboratory cultures. Indeed, the mechanism of initiation of lipid radical oxidation, which has been debated for many years, seems to be the homolytic cleavage of photochemically produced hydroperoxides in phytodetritus (Rontani et al., 2003b). Redox-active metal ions are generally considered as the initiators of perhaps greatest importance for lipid oxidation in biological systems (Pokorny, 1987; Schaich, 1992). They may direct the cleavage of hydroperoxides either through alkoxyl or peroxyl radicals. In classical culture media (such as f/2) the metal chelator EDTA, which is present in high amounts, tightly binds free catalytic metal ions and thus renders them unavailable. EDTA thus acts in the culture media as an antioxidant and strongly limits radical oxidation processes.

Recently, autoxidative damages in cells of *E. huxleyi* strain CS-57 could be induced after incubation of this strain under an atmosphere of air + 0.5% CO_2 (Rontani et al., 2007a). The presence of additional CO_2 allowed: (i) to induce a stress that favoured oxidative damage and (ii) to decrease the pH of the culture medium releasing metal ions from EDTA complexes, which can act as catalysts of hydroperoxide homolysis.

It was also demonstrated recently that viral infection (Evans et al., 2006) and autocatalytic programmed cell death (Bidle and Falkowski, 2004) of phytoplanktonic cells could also lead to elevated production of reactive oxygen species (ROS) able to induce the degradation of cell components.

3.1 Chlorophyll phytyl side-chain

Autoxidation of the esterified chlorophyll phytyl chain involves either addition of peroxyl radicals to the double bond or hydrogen abstraction at the allylic carbon 4 (Rontani and Aubert, 1994; Rontani and Aubert, 2005). Classical addition of peroxyl radical to the double bond gives a tertiary radical (Fig. 8). This radical can then: (i) lead to Z and E epoxides (**18 and 19**) by fast intramolecular homolytic substitution (Fossey et al., 1995), or (ii) react with molecular oxygen affording (after hydrogen abstraction on another molecule of substrate) a diperoxide (**20**) (Fig. 8). Subsequent $NaBH_4$-reduction and alkaline hydrolysis of these compounds gives 3,7,11,15-tetramethylhexadecan-1,2,3-triol (**21**) (Fig. 8). In contrast, abstraction (by photochemically-produced peroxyl radicals) of a hydrogen atom at the allylic carbon 4 of the phytyl chain and subsequent oxidation of the allylic radicals thus formed affords (after $NaBH_4$-reduction and alkaline hydrolysis) Z and E 3,7,11,15-tetramethylhexadec-3-en-1,2-diols (**3 and 4**) and Z and E 3,7,11,15-tetramethyl-hexadec-2-en-1,4-diols (**22 and 23**) (Fig. 8). Compounds **22 and 23** (which are well specific markers of free radical oxidation) could

•e detected in particulate matter samples (Marchand et al., 2005) and *E. huxleyi* cells (Rontani t al., 2007a) attesting to the involvement of such processes in senescent phytoplanktonic cells.

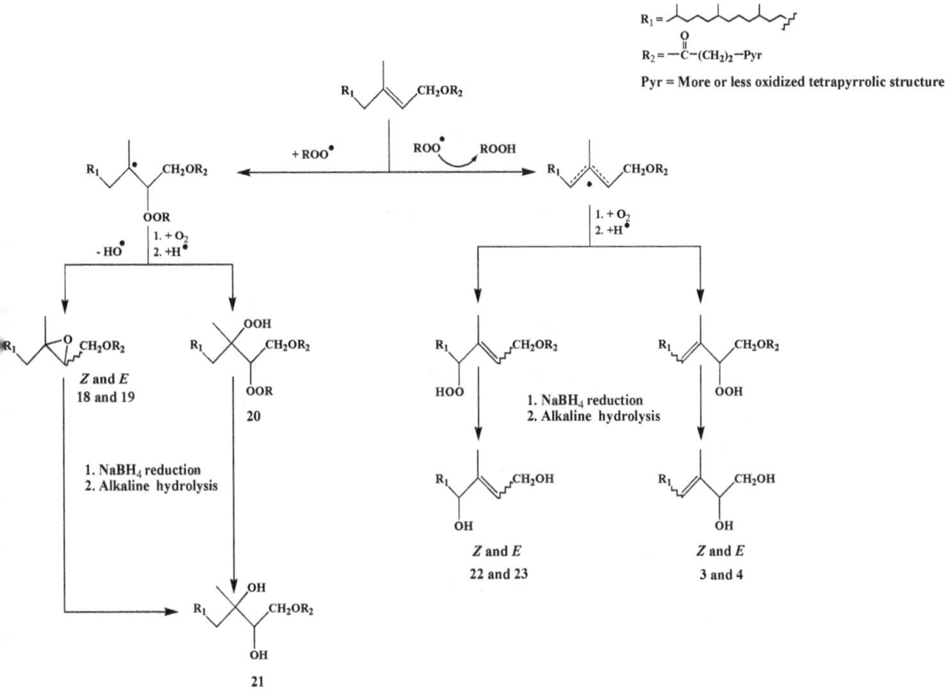

Fig. 8. Free radical-mediated oxidation of chlorophyll phytyl side-chain.

Free radical oxidation of chlorophyll phytyl chain appeared to be different in senescent cells of *S. costatum* (Rontani et al., 2003b). The differences observed were attributed to the well documented high chlorophyllase activity of this strain (Jeffrey and Hallegraeff, 1987) catalysing the hydrolysis of chlorophyll to free phytol and chlorophyllide. Indeed, in the case of free allylic alcohols hydrogen abstraction at carbon 1 is strongly favoured to the detriment of addition reactions (Huyser and Johnson, 1968).

3.2 Unsaturated fatty acids

Free radical oxidation of isolated classical 1,2-disubstituted double bonds generally involved mainly allylic hydrogen abstraction. Addition of peroxyl or alkoxyl radicals to the double bond becomes competitive only in the case of conjugated, terminal, or trisubstituted double bonds (Schaich, 2005). Effectively, autoxidation of mono-unsaturated fatty acids appears to mainly involve allylic hydrogen abstraction and subsequent oxidation of the allylic radical thus formed. For example, autoxidation of oleic acid mainly results in the formation of 9-hydroperoxyoctadec-*trans*-10-enoic (**24**), 10-hydroperoxyoctadec-*trans*-8-enoic (**25**), 11-hydroperoxyoctadec-*trans*-9-enoic (**26**), 11-hydroperoxyoctadec-*cis*-9-enoic (**27**), 8-hydroperoxyoctadec-*trans*-9-enoic (**28**) and 8-hydroperoxyoctadec-*cis*-9-enoic (**29**) acids (Fig. 9) (Frankel, 1998).

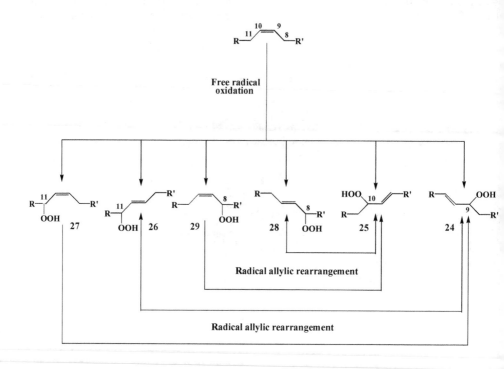

R = -(CH$_2$)$_6$-CH$_3$

R' = -(CH$_2$)$_6$-COOH

Fig. 9. Free radical-mediated oxidation of oleic acid.

Free radical oxidative processes can be easily characterised based on the presence of *cis* allylic hydroperoxyacids, which cannot be produced photochemically (see Fig. 4) and are specific products of these degradation processes (Porter et al., 1995; Frankel, 1998).

Large amounts of oxidation products of oleic acid could be detected in cells of *E. huxleyi* grown under an atmosphere of air + 0.5% CO$_2$ for 10 days (Rontani et al., 2007a). The presence (after NaBH$_4$-reduction) of a high proportion of 11-hydroxyoctadec-*cis*-9-enoic (**27**) and 8-hydroxyoctadec-*cis*-9-enoic (**29**) acids (Fig. 10) showed that under these conditions the degradation of oleic acid mainly involved free radical oxidation processes.

3.3 Δ^5-sterols

Free radical autoxidation of Δ^5-stenols yields mainly 7α- and 7β-hydroperoxides and, to a lesser extent, 5α/β,6α/β-epoxysterols and 3β,5α,6β-trihydroxysterols (Smith, 1981; Morrissey and Kiely, 2006) (Fig. 11).

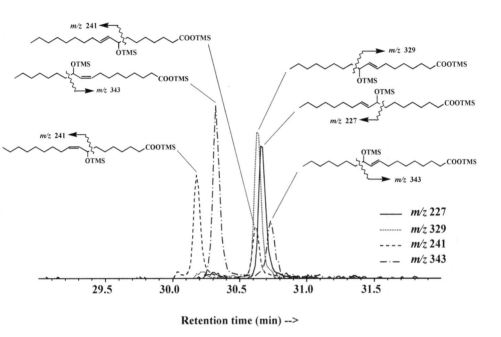

Fig. 10. Partial mass chromatogram of m/z 227, 329, 241 and 343 revealing the presence of oxidation products of oleic acid in the saponified fraction of E. *huxleyi* strain CS-57 grown under an atmosphere of air + 0.5% CO_2.

Owing to: their lack of specificity (possible formation by allylic rearrangement of photochemically-produced 5-hydroperoxides (see chapter 2.1.3), 7-hydroperoxides cannot be employed as tracers of autoxidation processes in phytodetritus. In contrast, it is generally considered that 5α/β,6α/β-epoxysterols arise mainly from peroxidation processes (Breuer and Björkhem, 1995; Giuffrida et al., 2004). Unfortunately, these compounds are not very stable and may be easily hydrolysed to the corresponding triol in seawater and during the treatment of the samples. 5α/β,6α/β-Epoxysterols and the corresponding 3β,5α,6β-trihydroxysterols were thus finally selected as tracers of sterol autoxidation..

5α/β,6α/β-Epoxysterols and 3β,5α,6β-trihydroxysterols corresponding to sitosterol, stigmasterol and campesterol were previously detected in young and old cell cultures of *Chenopodium rubrum* (Meyer and Spiteller, 1997). The results showed that the increase of these oxidation products well correlated with the age of the culture.

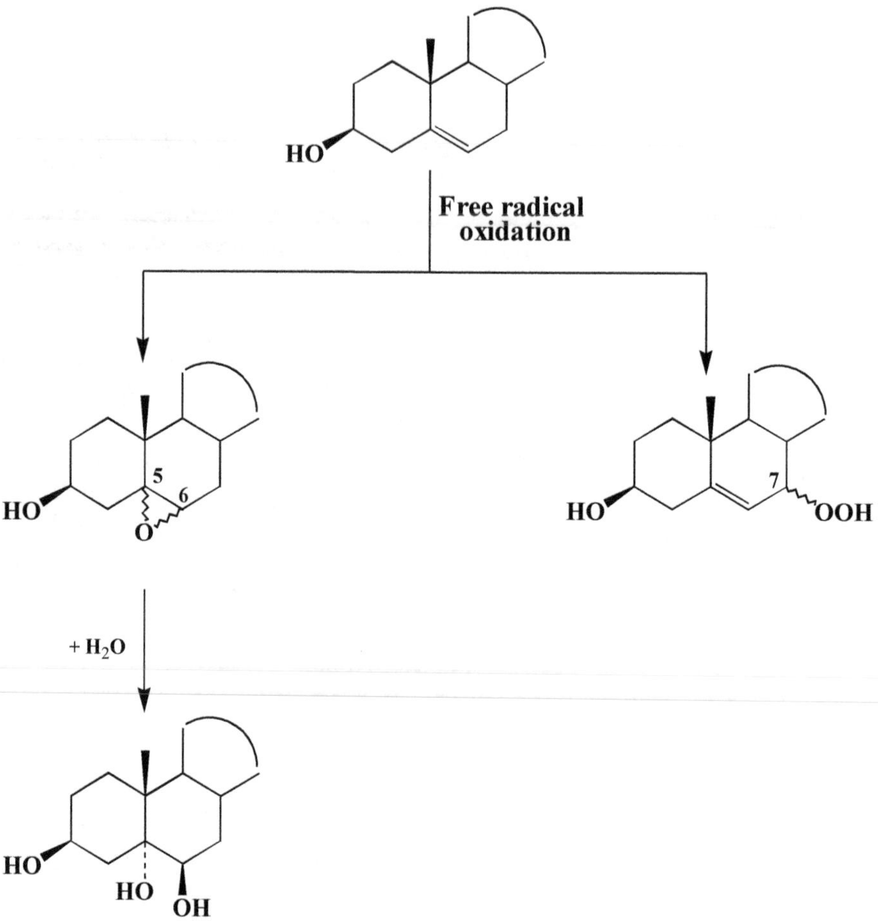

Fig. 11. Free radical-mediated oxidation of Δ^5 sterols.

3.4 Vitamin E

Vitamin E is relatively abundant in most photosynthetic organisms, such as higher plants (Rise et al., 1988; Schultz, 1990), cyanobacteria (Dasilva and Jensen, 1971), microalgae (Brown et al., 1999) and macroalgae (Sanchez-Machado et al., 2002), where it plays an essential role in the removal of toxic forms of oxygen (singlet oxygen, superoxide anion, hydroxyl and peroxyl radicals), by acting as sacrificial chemical scavenger (Halliwell, 1987); the process results in the irreversible oxidation of the tocopherol molecule. Vitamin E reacts rapidly with peroxyl radicals, affording small amounts of phytone (1), 4,8,12,16-tetramethylheptadecan-4-olide, α-tocopherylquinone and epoxy-α-tocopherylquinones, and dimers and trimers as major oxidation products (Liebler, 1994; Frankel, 1998; Rontani et al., 2007b) (Fig. 12).

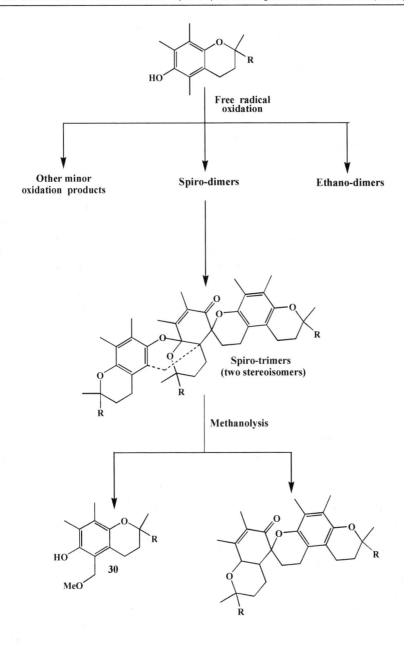

Fig. 12. Autoxidation of vitamin E and methanolysis of the foregoing trimers.

Isomeric trimers have been previously observed as products in numerous oxidations of vitamin E (e.g. Suarna et al., 1988; Krol et al., 2001). Such compounds cannot be easily detected since they are too heavy to be amenable by gas chromatography. However, methanolysis of the residues obtained resulted to the formation of high amounts of 5a-methoxytocopherol (30) arising from the methanolysis of the ketal group of trimers (Fig. 12) (Yamauchi et al., 1988). ESI-TOF MS analyses of oxidation products were also carried out in order to confirm the presence of high proportions of trimers (Nassiry et al., 2009).

Despite the intensive study of vitamin E oxidation since several decades, trimeric oxidation products could be detected in plants only very recently by Row et al. (2007). These authors detected these trimers in seeds of *Euryale ferox* containing extraordinarily high content of tocopherols. It is interesting to note that trimers were previously obtained as the major reaction products of vitamin E autoxidized under mild conditions in solution (1%) in methyl linoleate (Yamauchi et al., 1988). In plastoglobules, which are lipid monolayer subcompartments of the thylakoid membranes of chloroplasts (Maeda and Dellapenna, 2007), the concentration of tocopherols can reach 10% of the total fatty acids (Vidi et al., 2006). At such a concentration, the formation of a high proportion of trimers during photodynamic damages is thus very likely. In order to check this hypothesis, we searched for the presence of 5a-methoxytocopherol (30) after methanolysis of NaBH₄-reduced and non-reduced lipid extracts obtained from cells of *Emiliania huxleyi* strain TWP1 and *Chrysotila lamellosa* strain HAP17. The detection of significant amounts of this methanolysis product of trimers (Yamauchi et al., 1988) in these extracts (Nassiry et al., 2009) well supported the presence of such trimeric oxidation products of vitamin E in these algae.

3.5 Alkenones

The autoxidative reactivity of alkenones was studied in the laboratory in the presence of a radical initiator (di-*tert*-butyl nitroxide) and a radical enhancer (*tert*-butyl hydroperoxide) (Rontani et al., 2006). Alkenones appeared to be more sensitive towards oxidative free radical processes than analogues of other common marine lipids such as phytyl acetate, methyl oleate and cholesteryl acetate, and their oxidation rates increase in proportion with their number of double bonds. As the result of this increasing reactivity with degree of unsaturation, the $U_{37}^{K'}$ ratio increased significantly (up to 0.20) during the incubation.

Autoxidation of alkenones appears to mainly involve allylic hydrogen abstraction and subsequent oxidation of the allylic radical thus formed (Fig. 13). According to these processes, oxidation of each double bond of alkenones and subsequent NaBH₄ reduction affords four positional isomeric alkenediols. These compounds could be very useful indicators of autoxidation of alkenones but, unfortunately, they did not accumulate during the incubation. Indeed, due to the presence of additional reactive double bonds, hydroperoxyalkenones may undergo subsequent oxidation reactions affording, di-, tri- and tetrahydroperoxyalkenones according to the degree of unsaturation of the starting alkenone. In seawater, these different hydroperoxides may undergo two main degradative processes: (i) homolysis of the O-O bond leading to carbonyl (dehydration), alcoholic (reduction) and fragmentation (β-scission) products (Rontani et al., 2007c) and (ii) heterolysis of the O-O bond leading to the formation of two carbonyl fragments (Hock cleavage), this proton-catalysed cleavage being initiated by migration of groups to positive oxygen (Frimer, 1979). Dimeric and oligomeric compounds cross-linked through either peroxide or ether linkages (Frankel, 1998) may also be formed during autoxidation of alkenones.

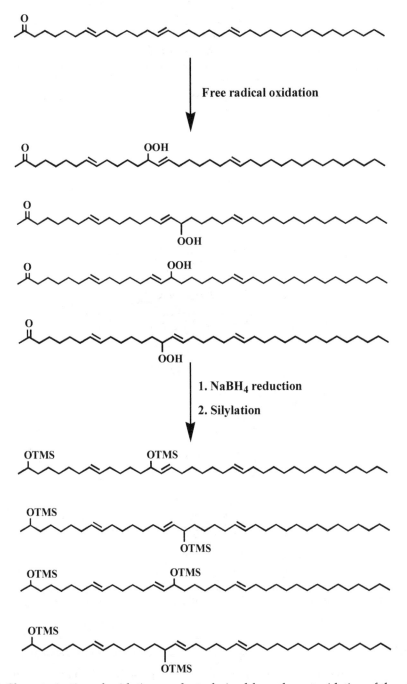

Fig. 13. Characterization of oxidation products derived from the autoxidation of the ω22 double bond of the $C_{37:3}$ alkenone (TMS = trimethylsilyl).

These results were corroborated by the further finding of significant amounts of alkenediols arising from NaBH$_4$-reduction of the corresponding hydroperoxyalkenones in cultures of *E. huxleyi* strain CS-57 grown under an atmosphere of air + 0.5% CO$_2$ (Rontani et al., 2007a) and more recently after incubation of a culture of the strain *E. huxleyi* TWP1 under darkness (Rontani, Unpublished results) (Fig. 14) both exhibiting an anomalously high unsaturation ratio It seems thus that autoxidation processes have the potential to affect alkenone distributions leading to a warm bias in estimates of palaeotemperatures derived from alkenone ratios in sediments.

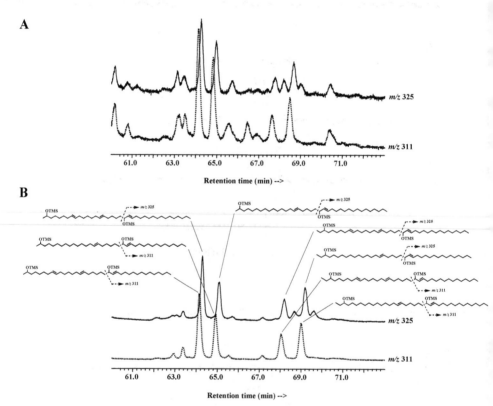

Fig. 14. Partial mass fragmentograms of *m/z* 311 and 325 revealing the presence of silylated C$_{37}$ and C$_{38}$ alkenediols after NaBH$_4$-reduction and silylation of the total lipid extract of *E. huxleyi* cells incubated under darkness (A) and standard autoxidation products of alkenones (B).

4. Conclusions

Due to the lack of adequate tracers, the role played by light-induced photochemical and free radical-mediated (autoxidative) processes during the degradation of lipid components of phototrophic organisms has been virtually ignored until now.

It was recently demonstrated that most of the unsaturated lipid components of these organisms (chlorophylls, carotenoids, unsaturated fatty acids, sterols, n-alkenes and HBI alkenes) could be photodegraded by visible and UV radiations during the senescence. This degradation mainly involves type II (i.e. involving $^1O_2$) photoprocesses. Singlet oxygen appeared to be sufficiently stable in this hydrophobic micro-environment to migrate outside the chloroplasts and affect the unsaturated components of cutins of higher plants.

Free radical-mediated oxidation (autoxidation) processes also intervene intensively during the senescence of phototrophic organisms. Induction of these processes seems to mainly result from the homolytic cleavage (catalyzed by some metal ions) of photochemically produced hydroperoxides. Unsaturated fatty acids, chlorophyll phytyl side-chain, vitamin E, sterols and alkenones appeared to be strongly affected by these degradative processes. In the case of alkenones, it is very important to note that autoxidative degradation processes may alter significantly their unsaturation ratio and thus constitute a potential source of biases during paleotemperature reconstruction.

5. Acknowledgements

Financial support over many years from the Centre National de la Recherche Scientifique (CNRS) and the Université de la Méditerranée is gratefully acknowledged.

6. References

Allard, W. G., Belt, S.T., Massé, G., Naumann, R., Robert, J.-M. & Rowland, S.J. (2001). Tetra-unsaturated sesterterpenoids (Haslenes) from *Haslea ostrearia* and related species. *Phytochemistry* 56, 795-800.

Belt, S.T., Allard, W.G., Massé, G., Robert, J.-M. & Rowland, S.J. (2000). Highly branched isoprenoids (HBIs): Identification of the most common and abundant sedimentary isomers. *Geochimica & Cosmochimica Acta* 64, 3839-3851.

Belt, S.T., Massé, G., Rowland, S.J., Poulin, M., Michel, C. & Leblanc, B. (2007). A novel chemical fossil of palaeo sea ice : IP$_{25}$. *Organic Geochemistry* 38, 16-27.

Belt, S.T., Vare, L.L., Massé, G., Manners, H., Price, J., MacLachlan, S., Andrews, J.T. & Schmidt, S. (2010). Striking similarities in temporal changes to seasonal sea ice conditions across the central Canadian Arctic Archipelago during the last 7,000 years. *Quaternary Science Reviews* 29, 3489-3504.

Bidle, K.D. & Falkowski, P.G. (2004). Cell death in planktonic, photosynthetic microorganisms. *Nature Review* 2, 643-655.

Boon, J.J., van der Meer, F.W., Schuyl, P.J., De Leeuw, J.W., Schenck, P.A. & Burlingame, A.L. (1978). Organic geochemical analyses of core samples from site 362 Walvis Ridge. DSDP Leg 40, *Initial Reports. DSDP* Leg 38, 39, 40 and 41, Suppl., pp. 627-637.

Breuer, O. & Blörkhem, I. (1995). Use of $^{18}O_2$ inhalation technique and mass isotopomer distribution analysis to study oxygenation of cholesterol in rat: evidence for in vitro formation of 7-oxo-, 7b-hydroxy-, and 25-hydroxycholesterol. *Journal of Biological Chemistry* 270, 20278-20284.

Brown, M.R., Mular, M., Miller, I., Farmer, C. & Trenerry, C. (1999). The vitamin content of microalgae used in aquaculture. *Journal of Applied Phycology* 11, 247-255.

Christodoulou S., Marty J.-C., Miquel J.-C., Volkman J.K. & Rontani J.-F. (2009). Use of lipids and their degradation products as biomarkers for carbon cycling in the northwestern Mediterranean Sea. *Marine Chemistry* 113, 25-40.

Christodoulou S., Joux F., Marty J.-C., Sempéré R. & Rontani J.-F. (2010). Comparative study of UV and visible light induced degradation of lipids in non-axenic senescent cells of *Emiliania huxleyi*. *Marine Chemistry* 119, 139-152.

Conte, M.H., Volkman, J.K. & Eglinton, G. (1994). Lipid biomarkers of the Haptophyta. In:Green, J.C., Leadbeater, B.S.C. (Eds.), *The Haptophyte Algae. Systematics Association* Special Volume No. 51, Clarendon Press, Oxford, pp. 351-377.

Cuny, P. & Rontani, J.-F. (1999). On the widespread occurrence of 3-methylidene-7,11,15-trimethylhexadecan-1,2-diol in the marine environment: a specific isoprenoid marker of chlorophyll photodegradation. *Marine Chemistry* 65, 155-165.

Cuny, P., Romano, J.-C., Beker, B. & Rontani, J.-F. (1999). Comparison of the photo-degradation rates of chlorophyll chlorin ring and phytol side chain in phytodetritus: is the phytyldiol versus phytol ratio (CPPI) a new biogeochemical index? *Journal of Experimental Marine Biology and Ecology* 237, 271-290.

Cuny, P., Marty, J.-C., Chiaverini, J., Vescovali, I., Raphel, D. & Rontani, J.-F. (2002). One-year seasonal survey of the chlorophyll photodegradation process in the Northwestern Mediterranean Sea. *Deep Sea Research II* 49, 1987-2005.

DaSilva, E.J. & Jensen, A. (1971). Content of α-tocopherol in some blue-green algae. *Biochimica et Biophysica Acta* 239, 345-347.

De Leeuw, J.W., van der Meer, J.W., Rijpstra, W.I.C. & Schenck, P.A. (1980). On the occurrence and structural identification of long chain ketones and hydrocarbons in sediments. In: Douglas, A.G., Maxwell, J.R. (Eds.), *Advances in Organic Geochemistry 1979*. Pergamon Press, Oxford, pp. 211-7.

Eltgroth, M.L., Watwood, R.L. & Wolfe, G.V. (2005). Production and cellular localization of neutral long-chain lipids in the Haptophyte algae *Isochrysis galbana* and *Emiliania huxleyi*. *Journal of Phycology* 41, 1000-1009.

Engel, N., Jenny, T. A., Mooser, V. & Gossauer, A. (1991). Chlorophyll catabolism in *Chlorella protothecoides*. Isolation and structure elucidation of a red biline derivative. *FEBS Letters* 293, 131-133.

Evans, C., Malin, G. & Mills, G.P. (2006). Viral infection of *Emilinia huxleyi* (Prymnesiophyceae) leads to elevated production of reactive oxygen species. *Journal of Phycology* 42, 1040-1047.

Foote, C.S. (1976). Photosensitized oxidation and singlet oxygen: consequences in biological systems. In: Pryor, W.A. (Ed.), *Free Radicals in Biology*. Academic Press, New York, pp. 85-133.

Fossey, J., Lefort, D. & Sorba, J. (1995). *Free radicals in organic chemistry*. Masson. Paris, pp 1-307.

Frankel, E.N., Neff, W.E. & Bessler, T.R. (1979). Analysis of autoxidized fats by gas chromatography-mass spectrometry: V. photosensitized oxidation. *Lipids* 14, 961-967.

Frankel, E.N. (1998). *Lipid Oxidation*. The Oily Press, Dundee.

Frimer, A.A. (1983). Singlet oxygen in peroxide chemistry. In: Patai, S. (Ed.), *The Chemistry of Functional Groups, Peroxides*. John Wiley & sons Ltd., pp. 202-229.

Gelin, F., Boogers, I., Noordeloos, A. A. M., Sinninghe Damsté, J. S., Riegman, R. & de Leeuw, J. W. (1997). Resistant biomacromolecules in marine microalgae of the classes Eustigmatophyceae and chlorophyceae: Geochemical implications. *Organic Geochemistry* 26, 659-675.

Giuffrida, F., Destaillats, F., Robert, F., Skibsted, L.H. & Dionisi F. (2004) Formation and hydrolysis of triacylclycerol and sterol epoxides : role of unsaturated triacylglycerol peroxyl radicals. *Free Radical Biology & medicine* 37, 104-114.

Grimalt, J.O., Rullkötter, J., Sicre, M.-A., Summons, R., Farrington, J., Harvey, H.R., Goñi, M. & Sawada, K. (2000). Modifications of the C_{37} alkenone and alkenoate composition in the water column and sediment: possible implications for sea surface temperature estimates in paleoceanography. *Geochemistry Geophysics and Geosystems* 1, doi: 10.1029/2000G000053.

Grossi, V., Raphel, D., Aubert, C. & Rontani, J.-F. (2000). The effect of growth temperature on the long-chain alkene composition in the marine coccolithophorid *Emiliania huxleyi*. *Phytochemistry* 54, 393-399.

Grossi, V., Beker, B., Geenevasen, J.A.J., Schouten, S., Raphel, D., Fontaine, M.-F. & Sinninghe-Damsté, J.S. (2004). C_{25} highly branched isoprenoid alkenes from the marine benthic diatom *Pleurosigma strigosum*. *Phytochemistry* 65, 3049-3055.

Halliwell, B. (1987). Oxidative damage, lipid peroxidation and antioxidant protection in chloroplasts. *Chemistry and Physics of lipids* 44, 327-340.

Harvey, H.R. (2000). Alteration processes of alkenones and related lipids in water columns and sediments. *Geochemistry Geophysics and Geosystems* 1, doi: 2000GC000054.

He, Y.Y. & Häder, D.P. (2002) Reactive oxygen species and UV-B effect on cyanobacteria. *Photochemical and Photobiological Sciences* 1, 729-736.

Heath, R.L. & Packer, L. (1968). Photoperoxidation in isolated chloroplasts. II. Role of electron transfer. *Archives of Biochemistry and Biophysics* 125, 850-857.

Hurst, J.R., Wilson, S.L. & Schuster, G.B. (1985). The ene reaction of singlet oxygen: kinetic and product evidence in support of a perepoxide intermediate. *Tetrahedron* 41, 2191-2197.

Huyser, E.S. & Johnson, K.L. (1968). Concerning the nature of the polar effect in hydrogen atom abstractions from alcohols, ethers and esters. *Journal of Organic Chemistry 33*, 3972-3974.

Iseo, S., Hyeon, S.B., Katsumura, S. & Sakan, T. (1972). Photooxygenation of carotenoids. II-The absolute configuration of loliolide and dihydroactinidiolide. *Tetrahedron Letters* 25, 2517-2520.

Iturraspe, J., Engel, N. & Gossauer, A. (1994). Chlorophyll catabolism. Isolation and structure elucidation of chlorophyll b catabolites in *Chlorella protothecoides*. *Phytochemistry* 35, 1387-1390.

Jaraula, C.M.B., Brassell, S.C, Morgan-Kiss, R, Doran, P.T. & Kenig, F. (2010) Origin and distribution of tri to pentaunsaturated alkenones in Lake Fryxell, East Antarctica. *Organic Geochemistry* 41, 386-397.

Jeffrey, S.W. & Hallegraeff, G.M. (1987). Chlorophyllase distribution in ten classes of phytoplankton: a problem for chlorophyll analysis. *Marine Ecology Progress Series* 35, 293-304.

Knox, J.P. & Dodge, A.D. (1985). Singlet oxygen and plants. *Phytochemistry* 24, 889-896.

Kolber, Z.S., Van Dover, C.L., Niederman, R.A. & Falkowski, P.G. (2000) Bacterial photosynthesis in surface waters of the open ocean. *Nature 407*, 177-179.

Kolber, Z.S., Plumley, F.G., Lang, A.S., Beatty, J.T., Blankenship, R.E., VanDover, C.L., Vetriani, C., Koblizek, M., Rathgeber, C. & Falkowski, P.G. (2001) Contribution of aerobic photoheterotrophic bacteria to the carbon cycle in the ocean. *Science 292*, 2492-2495.

Korytowski, W., Bachowski, G.J. & Girotti, A.W. (1992). Photoperoxidation of cholesterol in homogeneous solution, isolated membranes, and cells: comparison of the 5α- and 6β-hydroperoxides as indicators of singlet oxygen intermediacy. *Photochemistry and Photobiology 56*, 1-8.

Krol, E.S., Escalante, D.D.J. & Liebler, D.C. (2001). Mechanisms of dimer and trimer formation from ultraviolet-irradiated α-tocopherol. *Lipids 36*, 49-55.

Kulig, M.J. & Smith, L.L. (1973). Sterol metabolism. XXV. Cholesterol oxidation by singlet molecular oxygen. *Journal of Organic Chemistry 38*, 3639-3642.

Liebler, D.C. (1994). Tocopherone and epoxytocopherone products of vitamin E oxidation. *Methods in Enzymology 234*, 310-316.

MacIntosh, T.J. (1978). The effect of cholesterol on the structure of phosphatidylcholine bilayers. *Biochimica et Biophysica Acta 513*, 43-58.

Maeda, H. & DellaPenna, D. (2007). Tocopherol functions in photosynthetic organisms. *Current Opinion in Plant Biology 10*, 260-265.

Marchand, D. & Rontani, J.-F. (2001). Characterization of photooxidation and autoxidation products of phytoplanktonic monounsaturated fatty acids in marine particulate matter and recent sediments. *Organic Geochemistry 32*, 287-304.

Marchand, D. & Rontani, J.-F. (2003). Visible light-induced oxidation of lipid components of purple sulphur bacteria: A significant process in microbial mats, *Organic Geochemistry 34*, 61-79.

Marchand, D., Marty, J.-C., Miquel, J.-C. & Rontani, J.-F. (2005). Lipids and their oxidation products as biomarkers for carbon cycling in the northwestern Mediterranean Sea: results from a sediment trap study. *Marine Chemistry 95*, 129-147.

Marlowe, I.T., Green, J.C., Neal, A.C., Brassell, S.C., Eglinton, G. & Course, P.A. (1984). Long chain (n-C_{37}–C_{39}) alkenones in the Prymnesiophyceae. Distribution of alkenones and other lipids and their taxonomic significance. *British Phycology Journal 19*, 203-216.

Massé, G., Rowland, S.J., Sicre, M.-A., Jacob, J., Jansen, E. & Belt, S.T. (2008). Abrupt climate changes for Iceland during the last millennium: Evidence from high resolution sea ice reconstructions. *Earth and Planetary Science Letters 269*, 565-569.

Merzlyak, M.N. & Hendry, G.A.F. (1994). Free radical metabolism, pigment degradation and lipid peroxidation in leaves during senescence. *Proceedings of the Royal Society of Edinburgh 102B*, 459-471.

Meyer, W. & Spiteler, G. (1997). Oxidized phytosterols increase by ageing in photoautotrophic cell cultures of *Chenopodium rubrum*. *Phytochemistry 45*, 297-302.

Morrissey, P. A. & Kiely, M. (2006). Oxysterols: formation and biological function. In: *Advanced Dairy Chemistry 3rd edition*, Vol. 2 Lipids, Fox. P.F. & McSweeney, P.L.H., eds., Spinger, New-York, pp. 641-674.

Mouzdahir A., Grossi, V., Bakkas, S. & Rontani, J.-F. (2001). Photodegradation of long-chain alkenes in senescent cells of *Emiliania huxleyi* and *Nannochloropsis salina*. *Phytochemistry* 56, 677-684.

Müller, P.J., Kirst, G., Ruhland, G., von Storch, I. & Rosell-Melé, A. (1998). Calibration of the alkenone paleotemperature index $U_{37}^{K'}$ based on core-tops from the eastern South Atlantic and global ocean (60°N-60°S). *Geochimica & Cosmochimica Acta* 62, 1757-1772.

Nassiry, M., Aubert, C., Mouzdahir, A. & Rontani, J.-F. (2009) Generation of isoprenoid compounds and notably of prist-1-ene through photo- and autoxidative degradation of vitamin E. *Organic Geochemistry* 40, 38-50.

Neff, W.E., Frankel, E.N. & Fujimoto, K. (1988). Autoxidative dimerization of methyl linolenate and its monohydroperoxides, hydroperoxy epidioxides and dihydroperoxides. *Journal of the American Oil Chemical Society* 65, 616-623.

Nelson, J.R. (1993). Rates and possible mechanism of light-dependent degradation of pigments in detritus derived from phytoplankton. *Journal of Marine Research* 51, 155- 179.

Nickon, A. & Bagli, J.F. (1961). Reactivity and geochemistry in allylic systems. I. Stereochemistry of photosensitized oxygenation of monoolefins. *Journal of the American Chemical Society* 83, 1498-1508.

Pokorny, J. (1987). Major factors affecting the autoxidation of lipids. In: Chan, H.W.-S. (Ed.), *Autoxidation of Unsaturated Lipids*. Academic Press, London, pp. 141–206.

Porter, N.A., Caldwell, S.E. & Mills, K.A. (1995). Mechanisms of free radical oxidation of unsaturated lipids. *Lipids* 30, 277-290.

Prahl, F.G. & Wakeham, S.G. (1987). Calibration of unsaturation patterns in long-chain ketone compositions for palaeotemperature assessment. *Nature* 330, 367-369.

Prahl, F.G., Muehlhausen, L.A. & Zahnle, D.L. (1988). Further evaluation of long-chain alkenones as indicators of paleoceanographic conditions. *Geochimica & Cosmochimica Acta* 52, 2303-2310.

Prahl F. G., Rontani J.-F., Volkman J. K., Sparrow M. A. & Royer I. M. (2006). Unusual C₃₅ and C₃₆ alkenones in a paleoceanographic benchmark strain of *Emiliania huxleyi*. *Geochimica & Cosmochimica Acta* 70, 2856-2867.

Rechka, J.A. & Maxwell, J.R. (1988). Unusual long chain ketones of algal origin. *Tetrahedron Leters*. 29, 2599-600.

Repeta, D.J. (1989). Carotenoid diagenesis in recent marine sediments. II- Degradation of fucoxanthin to loliolide. *Geochimica & Cosmochimica Acta* 53, 699-707.

Rieley, G., Teece, M. A., Peakman, T. M., Raven, A. M., Greene, K. J., Clarke, T. P., Murray, M., Leftley, J. W., Campbell, C. N., Harris, R. P., Parkes, R. J. & Maxwell, J. R. (1998). Long-chain alkenes of the Haptophytes *Isochrysis galbana* and *Emiliania huxleyi*. *Lipids* 33, 617-625.

Rise, M., Cojocaru, M., Gottlieb, H.E. & Goldschmidt, E.E. (1988). Accumulation of α-tocopherol in senescing organs as related to chlorophyll degradation. *Plant Physiology* 89, 1028-1030.

Rontani, J.-F. & Aubert, C. (1994) Effect of oxy-free radicals upon the phytyl chain during chlorophyll-*a* photodegradation. *Journal of Photochemistry and Photobiology*, A79, 167-172.

Rontani, J.-F., Grossi, V., Faure, R. & Aubert, C. (1994). "Bound" 3-methylidene-7,11,15-trimethylhexadecan-1,2-diol: a new isoprenoid marker for the photodegradation of chlorophyll-a in seawater. *Organic Geochemistry* 21, 135-142.

Rontani, J.-F., Beker, B., Raphel, D. & Baillet, G. (1995). Photodegradation of chlorophyll phytyl chain in dead phytoplanktonic cells. *Journal of Photochemistry and Photobiology* 85A, 137-142.

Rontani, J.-F., Raphel, D. & Cuny, P. (1996a). Early diagenesis of the intact and photooxidized chlorophyll phytyl chain in a recent temperate sediment. *Organic Geochemistry* 24, 825-832.

Rontani, J.-F., Cuny, P. & Grossi, V. (1996b). Photodegradation of chlorophyll phytyl chain in senescent leaves of higher plants. *Phytochemistry* 42, 347-351.

Rontani, J.-F., Cuny, P. & Aubert, C. (1997a). Rates and mechanism of light-dependent degradation of sterols in senescing cells of phytoplankton. *Journal of Photochemistry and Photobiology* 111A, 139-144.

Rontani, J.-F., Cuny, P., Grossi, V. & Beker, B. (1997b). Stability of long-chain alkenones in senescing cells of *Emiliania huxleyi*: effect of photochemical and aerobic microbial degradation on the alkenone unsaturation ratio ($U^{K'}_{37}$). *Organic Geochemistry* 26, 503-509.

Rontani, J.-F., Cuny, P. & Grossi, V. (1998). Identification of a pool of lipid photoproducts in senescent phytoplanktonic cells. *Organic Geochemistry* 29, 1215-1225.

Rontani, J.-F. & Marchand, D. (2000) Δ⁵-Stenol photoproducts of phytoplanktonic origin: a potential source of hydroperoxides in marine sediments? *Organic Geochemistry* 31, 169-180.

Rontani J.-F., Koblizek M., Beker B., Bonin P. & Kolber Z. (2003a). On the origin of cis-vaccenic photodegradation products in the marine environment. *Lipids* 38, 1085-1092.

Rontani, J.-F., Rabourdin, A., Marchand, D. & Aubert, C. (2003b). Photochemical oxidation and autoxidation of chlorophyll phytyl side chain in senescent phytoplanktonic cells: potential sources of several acyclic isoprenoid compounds in the marine environment. *Lipids* 38, 241-253.

Rontani, J.-F. & Aubert, C. (2005). Characterization of isomeric allylic diols resulting from chlorophyll phytyl side-chain photo- and autoxidation by electron ionization gas chromatography/mass spectrometry. *Rapid Communications in Mass Spectrometry* 19, 637-646.

Rontani, J.-F., Rabourdin, A., Pinot, F., Kandel, S. & Aubert, C. (2005) Visible light-induced oxidation of unsaturated components of cutins: a significant process during the senescence of higher plants. *Phytochemistry* 66, 313-321.

Rontani, J.-F., Marty, J.-C., Miquel, J.-C. & Volkman, J.K. (2006). Free radical oxidation (autoxidation) of alkenones and other microalgal lipids in seawater. *Organic Geochemistry* 37, 354-368.

Rontani, J.-F., Jameson, I., Christodoulou, S. & Volkman, J.K. (2007a). Free radical oxidation (autoxidation) of alkenones and other lipids in cells of *Emiliania huxleyi*. *Phytochemistry* 68, 913-924.

Rontani, J.-F., Nassiry, M. & Mouzdahir, A. (2007b). Free radical oxidation (autoxidation) of α-tocopherol (vitamin E): A potential source of 4,8,12,16-tetramethylheptadecan-4-olide in the environment. *Organic Geochemistry* 38, 37-47.

Rontani, J;-F., Harji, R. & Volkman, J.K. (2007c) Biomarkers derived from heterolytic and homolytic cleavage of allylic hydroperoxides resulting from alkenone autoxidation. *Marine Chemistry* 107, 230-243.

Rontani, J.-F., Zabeti, N. & Wakeham, S.G. (2011a) Degradation of particulate organic matter in the equatorial Pacific Ocean: biotic or abiotic? *Limnology and Oceanography* 56, 333-349.

Rontani, J.-F., Belt, S.T., Vaultier, F. & Brown, T.A. (2011b) Visible light-induced photo-oxidation of highly branched isoprenoid (HBI) alkenes: a significant dependence on the number and nature of the double bonds. *Organic Geochemistry* 42, 812-822

Rosell-Melé, A., Carter, J.F., Parry, A.T. & Eglinton, G. (1995). Determination of the $U_{37}^{K'}$ index in geological samples. *Analytical Chemistry* 67, 1283-1289.

Row, L.-C., Ho, J.-C. & Chen, C.-M. (2007). Cerebrosides and tocopherol trimers from the seeds of *Euryale ferox*. *Journal of Natural products* 70, 1214-1217.

Rowland, S.J. & Robson, J.N. (1990). The widespread occurrence of highly branched isoprenoid acyclic C_{20}, C_{25} and C_{30} hydrocarbons in recent sediments and biota.- a review. *Marine Environmental Research* 30, 191-216.

Sanchez-Machado, D.I., Lopez-Hernandez, J. & Paseiro-Losada, P. (2002). High-performance liquid chromatographic determination of α-tocopherol in macroalgae. *Journal of Chromatography* 976A, 277-284.

Schaich, K.M. (1992). Metals and lipid oxidation. Contemporary issues. *Lipids* 27, 209–218.

Schaich, K.M. (2005) Lipid oxidation: theoretical aspects. In: *Bailey's Industrial Oil and Fat Products*, Sixth edition, Shahidi, F. (Ed.), John Wiley & Sons, pp. 269-355.

Schulte-Elte, K.H., Muller, B.L. & Pamingle, H. (1979). Photooxygenation of 3,3-dialkylsubstituted allyl alcohols. Occurrence of syn preference in the ene addition of $^1O_2$ at E/Z-isomeric alcohols. *Helvetica Chimica Acta* 62, 816–828.

Schultz, G. (1990). Biosynthesis of α-tocopherol in chloroplasts of higher plants. *Fett Wissenschaft Technologie* 92, 86-91.

Sinninghe-Damté, J.S., Muyzer, G., Abbas, B., Rampen, S.W., Massé, G., Allard, W.G., Belt, S.T., Robert, J.-M., Rowland, S.J., Moldovan, J.M., Barbanti, S.M., Fago, F.J., Denisevich, P., Dahl, J., Trindade, L.A.F. & Schouten, S. (2004). The rise of the Rhizosolenoid diatoms. *Science* 304, 584-587.

Smith, L.L. (1981). *The Autoxidation of Cholesterol*. Plenum Press, New York.

Suarna, C., Craig, D.C., Cross, K.J. & Southwell-Keely, P.T. (1988). Oxidations of vitamin E (α-tocopherol) and its model compound 2,2,5,7,8-pentamethyl-6-hydroxychroman. A new dimer. *Journal of Organic Chemistry* 53, 1281-1284.

Suwa, K., Kimura, T. & Schaap, A.P. (1977). Reactivity of singlet molecular oxygen with cholesterol in a phospholipidic membrane matrix: a model for oxidative damage of membranes. *Biochemistry and Biophysical Research Communications* 75, 785-792.

Vidi, P.-A., Kanwischer, M., Bagninsky, S., Austin, J.R., Csucs, G., Dörmann, P., Kessler, F. & Bréhélin, C. (2006). Tocopherol cyclase (VTE1) localization and vitamin E accumulation in chloroplast plastoglobule lipoprotein particles. *Journal of Biological Chemistry* 281, 11225-11234.

Volkman, J.K., Eglinton, G., Corner, E.D.S. & Forsberg, T.E.V. (1980). Long chain alkenes and alkenones in the marine coccolithophorid *Emiliania huxleyi*. *Phytochemistry* 19, 2619–2622.

Volkman, J. K., Barrett, S. M. & Dunstan, G. A. (1994). C_{25} and C_{30} highly branched isoprenoid alkenes in laboratory cultures of two marine diatoms. *Organic Geochemistry* 21, 407-414.

Volkman, J.K., Barrett, S.M., Blackburn, S.I. & Sikes, E.L. (1995). Alkenones in *Gephyrocapsa oceanica* - implications for studies of paleoclimate. *Geochimica & Cosmochimica Acta* 59, 513–520.

Wood, B.J.B. (1974). Fatty acids and saponifiable lipids. In: Steward, W.D. (Ed.), *Algal Physiology and Biochemistry*. University of California Press, Berkeley, pp. 236-265.

Yamauchi, R., Kato, K. & Ueno, Y. (1988). Formation of trimers of α-tocopherol and its model compound, 2,2,5,7,8-pentamethylchroman-6-ol, in autoxidizing methyl linoleate. *Lipids* 23, 779-783.

Yurkov, V.V. & Beatty, J.T. (1998) Aerobic anoxygenic phototrophic bacteria, *Microbiology and Molecular Biology Review* 62, 695-724.

Zafiriou, O.C. (1977) Marine organic photochemistry previewed. *Marine Chemistry* 5, 497-522.

Zafiriou, O.C., Joussot-Dubien, J., Zepp, R.G. & Zika, R.G. (1984) Photochemistry in natural waters. *Environmental Science and Technology* 18, 358A-370A.

4

Metabolic Regulation of Leaf Senescence in Sunflower (*Helianthus annuus* L.) Plants

Eloísa Agüera, Purificación Cabello,
Lourdes de la Mata, Estefanía Molina and Purificación de la Haba
Department of Botany, Ecology and Plant Physiology,
University of Córdoba
Spain

1. Introduction

The leaf is the main photosynthetic organ of plants and its development a complex process governed by a combination of environmental factors and intrinsic and genetically regulated signals (Van Lijsebettens & Clarke, 1998). Usually, leaf ontogeny includes an early phase of increasing photosynthetic rates while the leaf is actively expanding, a mature phase where such rates peak and a senescence phase where they decline (Gepstein, 1988; Miller et al., 2000). During early development, the leaf is a sink receiving nutrients from the rest of the plant; however, as soon as it reaches full photosynthetic capacity, it becomes the main source organ of the plant. After this productive period, the leaf enters the senescence phase, during which most compounds present in it are removed and reused (Hörtensteiner & Feller, 2002; Buchanan-Wollaston et al., 2003a). Leaf senescence, which is last stage in leaf development, is a highly regulated and programmed degeneration process governed by a variety of developmental and environmental signals (Lim et al., 2003). This important phase in the leaf lifespan period may last as long as leaf maturation and involves a shift from nutrient assimilation to nutrient remobilization and recycling (Guiboileau et al., 2010). In senescent leaf metabolism, carbon and nitrogen assimilation are replaced by catabolism of chlorophyll and macromolecules such as proteins, RNA and membrane lipids, the degradation of which marks the senescence phase. Unsurprisingly, senescence alters the expression of many genes. These senescence-associated genes include regulatory genes encoding transcription factors; genes involved in degradative processes that code for hydrolytic enzymes such as proteases, lipases and ribonucleases; and genes with secondary functions in senescence that code for proteins involved in nutrient remobilization (e.g. glutamine synthetase, which catalyses the conversion of ammonium into glutamine to enable nitrogen recycling in senescing cells) (Taiz & Zeiger, 2010). Environmental cues such as day length and temperature, and various biotic and abiotic sources of stress, can also affect the initiation and progress of such a high complex as leaf senescence.

During senescence, some metabolic pathways are triggered and others turned off. These dramatic metabolic changes result in orderly degradation of cellular structures, starting with chloroplasts (Wiedemuth et al., 2005), and also in the subsequent remobilization of the resulting materials. Chloroplasts play a dual role; thus, they are the main source of nitrogen and also the regulators of their own degradation during senescence (Zapata et al., 2005). Most of the protein in green cells is located in chloroplasts, which thus constitute their main reserves of organic nitrogen. Efficient recycling of nitrogen from the photosynthetic apparatus during early senescence requires the presence of intact mitochondrial, nuclear and cellular membranes (Gan & Amasino, 1997; Nam, 1997; Noodén et al., 1997; Hörtensteiner & Feller, 2002; Cabello et al., 2006). Leaf proteins (particularly photosynthetic proteins) are extensively degraded during senescence (Martínez et al., 2008), which confirms that one of the primary functions of leaf senescence is to recycle nutrients (especially through nitrogen remobilization) (Himelblau & Amasino, 2001). Protein breakdown starts early in senescence and proteolysis is believed to start within chloroplasts. Some proteins (e.g. chlorophyll-binding light-harvesting proteins LHCII) seem to be entirely degraded within chloroplasts, whereas Rubisco and other chloroplastic proteins may be broken down via a hybrid pathway involving both chloroplasts and extraplastidic compartments such as the central vacuole and small senescence-associated vacuoles (SAVs), which are absent from mature, non-senescing leaves but present in large numbers during senescence (Otegui et al., 2005; Martínez et al., 2008). Degradation of chloroplastic proteins releases potentially phototoxic chlorophylls that necessitate degradation. Therefore, leaf senescence is characterized by a decline in photosynthetic activity and chlorophyll content, and the rapid chlorophyll loss associated with chloroplast degeneration is frequently used as a biomarker for the start of senescence. Although chlorophyll degradation is an early senescence signal, leaf yellowing is not an appropriate marker of early senescence because it is observed when senescence has progressed to a great extent (Diaz et al., 2005). Nitrogen and carbon metabolism plays a crucial role in the senescence process, which is seemingly governed by both external and internal factors. Thus, leaf senescence induction involves the joint action of external (nitrogen availability, light) and internal signals (regulating metabolites, C/N ratio) (Wingler et al., 2006; Wingler & Roitsch, 2008).

Other important signals for induction or progression of senescence include the redox status of leaf cells and the production of reactive oxygen species (ROS) such as hydrogen peroxide and superoxide radical (Kukavica & Veljovic-Jovanovic, 2004; Zimmermann & Zentgraf, 2005). There are many sources of reactive oxygen species, which are produced during aerobic metabolism in chloroplasts, mitochondria and peroxisomes in both photosynthetically active and senescent cells. The toxicity of these reactive species is dictated by various enzymatic and non-enzymatic protective antioxidant defences. Superoxide dismutases, catalases, peroxidases and the ascorbate–glutathione cycle enzymes are the primary antioxidant enzymes. Plant ageing increases oxidative stress and the levels of reactive oxygen species, which may additionally diminish antioxidant protection (Buchanan-Wollaston et al., 2003b; Zimmermann & Zentgraf, 2005). Chloroplasts are probably the main target of age-associated oxidative stress in plants (Munné-Bosch & Alegre, 2002). Therefore, a plausible model for regulation of leaf senescence is a shifted balance between the production of reactive oxygen species and their removal by antioxidant systems.

In this chapter, we describe various aspects of leaf senescence in sunflower plants, with special emphasis on changes in the contents of some nitrogen and carbon metabolites potentially acting as regulators or markers of senescence during sunflower leaf development, and also on the role of oxidative stress in this process and the influence of external factors such nitrogen supply and irradiance exposition on it.

. Growth-related parameters and photosynthetic activity during sunflower eaf senescence

Ve examined various markers widely used to monitor leaf development (viz. hotosynthetic pigment level, protein content and CO_2 fixation rate) in primary leaves of unflower plants grown for 42 days. The start of senescence in sunflower plants was ssociated with a considerable decrease in protein content and specific leaf masses referred s weight (Table 1).

Leaf age (days)	Soluble protein (mg g^{-1} DW)	Specific leaf mass (mg DW cm^{-2})
16	152.3 ± 9.4	2.2 ± 0.11
22	178.5 ± 7.7	3.1 ± 0.28
28	108.1 ± 4.6	3.0 ± 0.27
36	89.6 ± 1.9	2.5 ± 0.23
42	62.2 ± 1.4	2.2 ± 0.29

'able 1. Changes in soluble protein and specific leaf mass during sunflower primary leaf geing. Data are means ± SD for duplicate determinations in three separated experiments.

'hese changes may reflect alterations in N and C compound distributions as a consequence f N remobilization, the efficiency of which is related to the ratio between biomass in the ink and source organs (Wiedemuth et al., 2005; Diaz et al., 2008). Since chloroplasts contain he largest amounts of protein in leaves, their breakdown releases most of the nitrogen that s reused by other plant organs. The mechanisms behind chloroplast degradation in enescing leaves are poorly understood (especially those for the degradation of Rubisco and hlorophyll-binding light-harvesting proteins, which are the most abundant chloroplastic roteins) (Martínez et al., 2008). Chloroplasts contain a large number of proteases, some of vhich are encoded by senescence-associated genes, which are up-regulated during enescence. Degradation of some thylakoid proteins such as LHCII seemingly occurs xclusively within chloroplasts and requires the prior release and breakdown of pigments Hörtensteiner & Feller, 2002; Buchanan-Wollaston et al., 2003a). CND41 protease is believed o be involved in Rubisco degradation and in the translocation of nitrogen during enescence in tobacco leaves (Kato et al., 2004, 2005). However, the central vacuole and ;AVs also play a role here, as they help complete the degradation of Rubisco and other tromal proteins (Martínez et al., 2008). The relative rates of degradation of some hotosynthetic components may be altered by the environmental conditions. Thus, LHCII legradation in rice is delayed by low irradiances (Hidema et al., 1991). Also, the protein ontent in senescing sunflower leaves was found to drop earlier in nitrogen-deficient plants han in high-nitrogen plants (Agüera et al., 2010). Changes in photosynthetic pigment ontents also indicate progress of leaf senescence (Yoo et al., 2003; Guo & Gan, 2005;)ugham et al., 2008). The chlorophyll breakdown pathways operating during leaf enescence are well-known and require pigment degradation and avoiding photodamage in rder to maintain the ability to export released nutrients to other plant parts (Hörtensteiner,

2006; Ougham et al., 2008). Chlorophylls in sunflower plants are more susceptible to degradation than are carotenoids during leaf senescence, and both total chlorophyll and carotenoid contents are high in young and mature leaves, their levels peaking at 22 days and decreasing afterwards during senescence (Fig. 1). Carotenoid degradation is usually slower than chlorophyll breakdown and can be especially complex depending on the particular pigment species (Suzuki & Shioi, 2004).

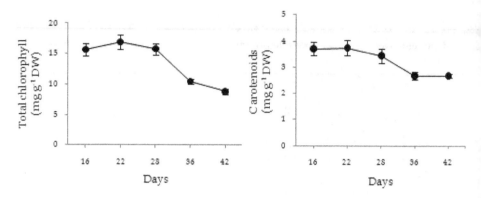

Fig. 1. Changes in pigment levels during ageing of sunflower primary leaves. Data are means ± SD for duplicate determinations in three separated experiments.

Chlorophyll loss in sunflower plants is also a typical phenomenon of leaf senescence of potential use as an indicator. The marked decrease in total chlorophyll observed after 28 days is mainly due to the loss of chlorophyll *a*, which is the form most strongly affected by leaf ageing as revealed by a significant decrease in Chl *a*/Chl *b* ratio in senescent leaves (Cabello et al., 2006). In radish cotyledons, however, the ratio of Chl *a* to Chl *b* increases slightly during senescence, which suggests that Chl *b* is degraded faster than is Chl *a* (Suzuki & Shioi, 2004).

Other typical changes observed during senescence are a rapid decline in photosynthetic activity, which may be a senescence-inducing signal (Bleecker & Patterson, 1997; Quirino et al., 2000), and a reduction in transpiration rate, which is probably due to an increase in abscisic acid levels inducing stomatal closure, although this is not a direct induction factor for senescence (Weaver & Amasino, 2001). A marked decrease in CO_2 fixation rate and transpiration in sunflower plants was observed during natural leaf senescence, a process that starts and develops in plants aged 28–42 days (Fig. 2).

Although natural senescence is the final stage of leaf development, it may start prematurely by effect of exposure to environmental stress or nutrient deprivation (Quirino et al., 2000; Lim et al., 2003, 2007, Wingler et al., 2009). In fact, poor nitrogen nutrition and exposure to high irradiance are known to lead to early senescence in sunflower leaves (Agüera et al., 2010). Thus, the decrease in chlorophyll content associated to leaf senescence starts earlier in sunflower plants grown with low nitrogen, which suggests that leaf senescence is accelerated under these conditions. In addition, the decline in photosynthetic activity is more apparent with nitrogen deficiency (Agüera et al., 2010). Similarly, the loss of photosynthetic activity is more marked in leaves of sunflower plants grown at high irradiance than in others grown at a low photon flux density, also indicating that an increased irradiance may accelerate leaf senescence.

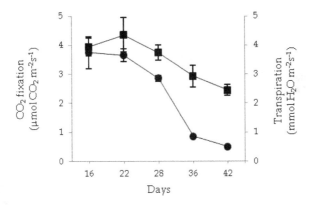

Fig. 2. Carbon dioxide fixation rates and transpiration in sunflower primary leaves of different age. Data are means ± SD of measured values on primary leaves of ten plants randomly selected for each age.

3. Carbon and nitrogen metabolites as regulators of leaf senescence in sunflower plants

The contents in soluble sugars of sunflower plants increase with leaf ageing, and the opposite holds for the starch content. Our results show that accumulation of soluble sugars in plants grown at high irradiance is not much greater than in plants grown at low irradiance, although a substantial increase in the monosaccharide-to-sucrose ratio is observed at the start of senescence (especially at high irradiance levels) (Fig. 3). The accumulation of soluble sugars is associated to leaf age but unrelated to photosynthetic activity because CO_2 fixation rates decrease during ageing; rather, it is due to starch hydrolysis. The increase in soluble sugars may also be ascribed to senescence causing a loss of functional and structural integrity in cell membranes, thereby boosting membrane lipid catabolism and hence sugar production by gluconeogenesis (Buchanan-Wollaston et al., 2003b; Lim et al., 2007). Leaf senescence is a plastic process triggered by a variety of external and internal factors (Weaver & Amasino, 2001; Buchanan-Wollaston et al., 2003a; Balibrea-Lara et al., 2004; Wingler et al., 2006). Senescence reduces photosynthetic carbon fixation, but is important for the recycling of nitrogen and other nutrients (Díaz et al., 2005; Wingler et al., 2005). By virtue of its lying at the crossroads of carbon and nitrogen metabolism, senescence is regulated by carbon and nitrogen signals. Increasing evidence suggests a role for hexose accumulation in ageing leaves as a signal for either senescence initiation or acceleration in annual plants (Masclaux et al., 2000; Moore et al., 2003; Díaz et al., 2005; Masclaux-Daubresse et al., 2005; Parrott et al., 2005; Pourtau et al., 2006; Wingler & Roitsch, 2008; Agüera et al., 2010). Recently, the role of sugar accumulation or starvation in leaf senescence was critically evaluated by van Doorn (2008), who pointed out that little is known about sugar concentrations and senescence regulation in different tissues and cells.

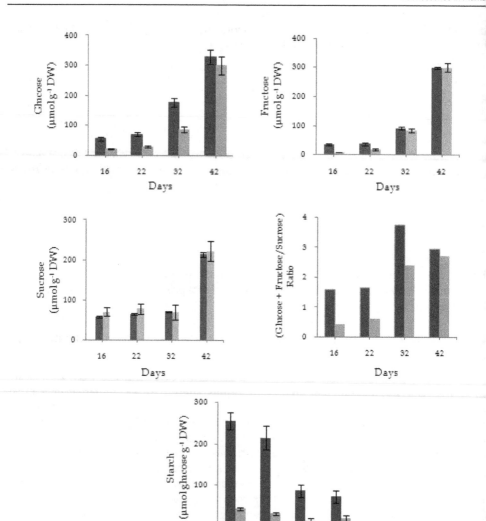

Fig. 3. Changes in glucose, fructose, sucrose and starch contents, and in hexoses-to-sucrose ratio, during development of sunflower primary leaves. Plants were grown at 125 μmol photons m^{-2} s^{-1} (grey bars) or 350 μmol photons m^{-2} s^{-1} (black bars). Data are means ± SD for duplicate determinations in three separate experiments.

Although sugars may not always be the direct cause of leaf senescence, there is enough evidence suggesting that sugar signalling plays a role in senescence regulation in a complex network involving a variety of other signals (Masclaux-Daubresse et al., 2007; Wingler & Roitsch, 2008; Wingler et al., 2009). Thus, cytokinin oxidase/dehydrogenase activity and senescence are positively correlated. The enzyme probably boosts senescence by destroying cytokinins and light is known to increase cytokinin oxidase/dehydrogenase activity during senescence of barley leaf segments (Schlüter et al., 2011).

Some results also suggest that leaf senescence is regulated by the carbon–nitrogen balance (Masclaux et al., 2000). However, in spite of the drastic changes in leaf metabolism occurring during senescence, carbon and nitrogen metabolite contents have scarcely been determined (Diaz et al., 2005). Cabello et al. (2006) found sunflower leaf senescence to be associated with significant changes in the contents of carbon and nitrogen metabolites. The highest ammonium concentrations were found in young and senescent leaves, as reported in tobacco (Masclaux et al., 2000). Our results indicate that sunflower plants exhibit their peak ammonium contents in young and late senescing leaves (Table 2). The high ammonium contents of young leaves are probably a result of strong photosynthetic nitrate reduction activity and photorespiration. In addition, young leaves have low contents in soluble carbohydrates, and sugar availability is known to be a limiting factor for ammonium assimilation (Morcuende et al., 1998). The high ammonium contents of senescent leaves are mainly due to protein degradation, amino acid deamination and nucleic acid catabolism, but also to photorespiration.

Senescent leaves contain low levels of free amino acids, probably because their remobilization is essential with a view to supplying developing organs in the plant (Buchanan-Wollaston, 1997). The concentrations of glutamate (a precursor of other amino acids) and aspartate (a direct product of glutamate transamination) decrease in the final stages of senescence in *Arabidopsis*. Glutamine and asparagine, the major amino acids translocated in the phloem sap, are mobilized more efficiently during late senescence (Diaz et al., 2005). As suggested by a genome array study (Lin & Wu, 2004), the synthesis of asparagine for nitrogen remobilization during dark-induced leaf senescence in *Arabidopsis* seems to occur via a novel biochemical pathway. Cabello et al. (2006) found glutamate to be the most abundant free amino acid in sunflower leaves as previously also found in rice (Kamachi et al., 1991), tobacco (Masclaux et al., 2000; Tercé-Laforgue et al., 2004) and *Arabidopsis* (Diaz et al., 2005). The ratio (Glu + Asp)/(Gln + Asn) peaked in sunflower leaves of 22 days, but decreased gradually in leaves of 28, 36 and 42 days (Table 2), which suggests that N-rich amino acids (specially Asn, which has a lower C to N ratio) are produced for efficient export from leaves in late senescence, as proposed for *Arabidopsis* (Diaz et al., 2005).

Leaf age (days)	Ammonium ($\mu mol \ g^{-1} DW$)	(Glu + Asp/Gln + Asn) Ratio
16	11.40 ± 1.0	2.11
22	8.57 ± 0.9	2.29
28	7.29 ± 0.7	1.77
36	8.94 ± 0.5	1.51
42	10.91 ± 0.7	1.44

Table 2. Changes in ammonium content and glutamate + aspartate / glutamine + asparagine ratio during sunflower primary leaf ageing. Data are means ± SD for duplicate determinations in three separated experiments

We examined changes in glutamine synthetase (GS) expression and activity during leaf development (Cabello et al., 2006). GS, which is the key enzyme in ammonia assimilation, is present as chloroplastic (GS2) and cytosolic (GS1) isoforms in sunflower leaves (Cabello et al., 1991). In order to confirm whether these isoforms are differently affected by senescence in sunflower leaves, we determined their specific activity during plant development. As shown in Figure 4, total GS activity decreased with the leaf age. The decrease was consequence of a strong decline in chloroplastic GS2 activity. On the other hand, cytosolic GS1 activity increased with ageing. It should be noted that GS1 was the predominant isoform in senescent leaves of 42 days, but accounted for only 7 % of total GS activity in young leaves (16 days). As a result, the GS2/GS1 ratio decreased from 13.3 in young leaves (16 days) to 0.9 in senescent leaves (42 days) (Fig. 4). These results indicate that leaf senescence has an adverse effect on the activity of chloroplastic GS2 (the main glutamine synthetase isoform) and reduces total GS activity despite its boosting GS1 activity.

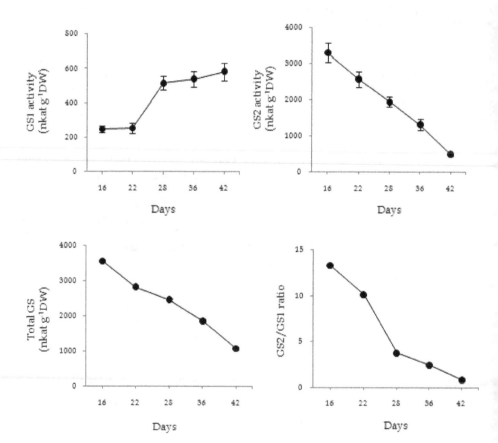

Fig. 4. Effect of ageing on total GS activity and on the activities of GS1 and GS2 isoforms in sunflower leaves. Data are means ± SD of duplicate determinations from three separated experiments.

Ageing affects glutamine synthetase activity but plays a direct role in the regulation of GS gene expression (Cabello et al., 2006). A Northern blot test using a probe corresponding to an internal fragment from *Helianthus annuus* GS2 cDNA revealed that the levels of GS2 transcripts decreased during leaf development and were very low in the late stage of senescence (42 days) (Fig. 5). Glutamine synthetase activity has been found to decrease during natural leaf senescence in a wide variety of plants including cereals, tomato and tobacco (Streit & Feller, 1983; Kamachi et al., 1991; Pérez-Rodríguez & Valpuesta, 1996; Masclaux et al., 2000). This loss of activity is mainly due to a gradual decrease in the major plastidial GS2 isoform since the cytosolic GS1 isoform remains constant or increases during leaf ageing (Pérez-Rodríguez & Valpuesta, 1996; Masclaux et al., 2000).

Northern blots and immunological analyses indicate that both GS transcripts and polypeptides are affected (Pérez-Rodríguez & Valpuesta 1996). GS1 plays a major role in the synthesis of glutamine for transport and remobilization of leaf organic nitrogen (Tercé-Laforgue et al., 2004), whereas GS2 takes part in the reassimilation of ammonium from photorespiration in photosynthetic tissues (Kamachi et al., 1992). The stimulation of the cytosolic GS1 isoform during senescence can be ascribed to the need for toxic ammonium to be reassimilated in order to produce glutamine for export to sink organs; this has led some authors to assume a shift in ammonia assimilation from the chloroplast to the cytosol of leaf cells during senescence (Brugière et al., 2000). Total GS activity was found to drop by a effect of a strong decrease in GS2 activity was found during sunflower leaf ageing despite the simultaneous increase in GS1 activity. GS2 transcript levels also diminished during ageing. Our results (Figs. 4 and 5) are therefore consistent with others previously reported for tomato and tobacco.

Time (days)

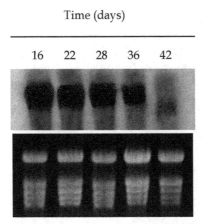

Fig. 5. Effect of ageing on GS2 mRNA accumulation in sunflower leaves.

Amino acids and other metabolites related to N metabolism deficit may act as signals to induce senescence in combination with hexose accumulation. Thus, leaf senescence in sunflower plants is induced by high sugar levels and accelerated by a low nitrogen supply, which supports the view that high sugar/low nitrogen conditions trigger senescence and facilitate its development (Wingler et al., 2009). Our results suggest that leaf senescence in sunflower plants is accelerated by nitrogen deficiency and high irradiance, and also that some factors such the levels of soluble sugars and amino acids may interact in a complex network to promote this process.

4. Oxidative stress in sunflower plants

Leaf senescence is an oxidative process that involves degradation of cellular and subcellular structures and macromolecules, and mobilization of the degradation products to other parts of the plants (Vanacker et al., 2006). Oxidative stress during senescence may be caused or increased by a loss of antioxidant enzymatic activities (Zimmermann & Zentgraf, 2005, Zimmermann et al., 2006; Procházkova & Wilhelmova, 2007). Senescence is also accompanied by an increase in ROS, one of the origins of which is an imbalance between the production and consumption of electrons in the photosynthetic electron transport chain caused by preferential inhibition of stromal reactions in contrast with photosystem II photochemistry (Špundová et al., 2003). The inhibition of stromal reactions increases the electron flow to molecular oxygen and causes ROS to accumulate and chloroplast components to be damaged as a result (Špundová et al., 2005; Couée et al., 2006). Chloroplasts are the main source of ROS in plants (Zimmermann & Zentgraf, 2005) and also the major target of oxidative damage (Munné-Bosch & Alegre, 2002). Stromal protein degradation during leaf senescence may be initiated by oxidative processes associated with the generation of free radicals and reactive species (Procházkova et al., 2001). Like Rubisco and other chloroplastic proteins, GS2 is susceptible to degradation initiated by reactive oxygen species (Ishida et al., 2002). The chloroplastic GS2 isoform is one of the first targets of oxidative damage at high irradiation levels (Palatnik et al., 1999). Oxidized GS becomes more susceptible to proteolysis (Ortega et al., 1999); under photo-oxidative stress, GS2 cleavage occurs preferentially around the catalytic site (Ishida et al., 2002). Senescence may therefore have a direct impact on GS2 activity through enzyme degradation initiated by reactive oxygen species as reported in Rubisco (Ishida et al., 1997; Roulin & Feller, 1998). Our results indicate that the decrease in GS2/GS1 ratio during sunflower leaf ageing may be partly due to a different sensitivity to oxidative stress of the two isoforms; in fact, chloroplastic GS2 is much more sensitive to oxidative modification *in vitro* than is cytosolic GS1 (Cabello et al., 2006). Therefore, ageing induces oxidative stress in sunflower leaves and can thus have an adverse effect on chloroplastic GS2, as well as on photosynthetic pigments. Antioxidant enzyme activities in sunflower leaves were found to decline during late senescence (42 days). Similar results have been reported for tobacco (Dhindsa et al., 1981), *Arabidopsis* (Ye et al., 2000), pea (Olsson, 1995) and maize (Procházkova et al., 2001). Oxidative stress during late senescence may be caused or increased by the loss of antioxidant enzymatic activities (Zimmermann & Zentgraf, 2005). Also, the decline in antioxidant activities is believed to be a consequence rather than the origin of senescence (Dertinger et al., 2003).

Susceptibility to oxidative stress depends on the overall balance between production of oxidants and cell antioxidant capability. In sunflower plants, considerable oxidative stress has been observed *in vivo* during leaf senescence, as revealed by lipid peroxidation, H_2O_2 accumulation and a decrease in the levels of antioxidant enzymes such as catalase, ascorbate peroxidase and superoxide dismutase (Table 3). Lipid peroxidation only occurs during the late stage of senescence (Berger et al., 2001; Jongebloed et al., 2004; Wingler et al., 2005). High irradiance causes reversible photoinhibition of photosynthesis in pea chloroplasts and increases ROS potentially regulating the accumulation of mRNA encoding antioxidant enzymes (Hernández et al., 2006).

Age	H_2O_2	Catalase	Ascorbate peroxidase	Superoxide dismutase	Lipid peroxidation
(days)	(μmol g^{-1} DW)		(U g^{-1} DW)		(nmol MDA g^{-1} DW)
16	1.22 ± 0.15	1.12 ± 0.10	17.21 ± 1.22	336.4 ± 28	87.6 ± 8.2
22	1.38 ± 0.14	1.70 ± 0.12	17.99 ± 2.12	356.2 ± 39	85.6 ± 7.4
28	3.84 ± 0.42	2.25 ± 0.26	28.22 ± 3.22	538.5 ± 42	155.5 ± 12.3
36	4.76 ± 0.30	1.94 ± 0.17	21.34 ± 2.19	1450.5 ± 112	171.5 ±14.5
42	5.28 ± 0.51	1.24 ± 0.15	14.53 ± 1.17	985.2 ± 92	188.4 ± 12.8

Table 3. Hydrogen peroxide accumulation, catalase, ascorbate peroxidase and superoxide dismutase activities, and lipid peroxidation levels during sunflower primary leaf development. Data are means ± SD for duplicate determinations in three separated experiments.

The activity and expression of antioxidant enzymes are seemingly sensitive to high irradiance stress (Yoshimura et al., 2000; Hernández et al., 2004).

We found H_2O_2 accumulation in senescent sunflower to be slightly more marked in plants grown under a nitrogen deficiency; the differences, however, were not large enough to assume that H_2O_2 is a major factor regulating the induction of leaf senescence in N-deficient plants (Table 3). Interestingly, catalase and ascorbate peroxidase activity decreased steadily in plants grown with low nitrogen, but increased during early leaf development and then declined during senescence in plants grown with high nitrogen (Agüera et al., 2010). Production of ROS during leaf senescence is essentially governed by chloroplasts, which have a strong photooxidative potential (Zapata et al., 2005). A simultaneous increase in lipid peroxidation was observed. Mutations in the *Arabidopsis CPR5/OLD1* gene may cause early senescence through deregulation of the cellular redox balance (Jing et al., 2008). Also, there is evidence suggesting that inadequate oxidant and carbonyl group production are intrinsically related to plant ageing, and that low mitochondrial, superoxide dismutase and ascorbate peroxidase activities may contribute to extensive protein carbonylation (Vanacker et al., 2006; Srivalli & Khanna-Chopra, 2009).

In conclusion, during sunflower leaf development some coordinated metabolic and physiological changes are produced, and the senescence process induces significant alterations in the levels of carbon and nitrogen metabolites. Glutamine synthetase of sunflower leaves is regulated both at transcriptional and enzyme levels during leaf ontogeny. Post-translational regulation of the GS2 isoform could be due, at least partially, to oxidative processes. GS activity may be used as a biochemical marker of leaf ageing, since the beginning of senescence at about 28 days is accompanied by a drastic drop in the GS2/GS1 ratio due to the increase of the cytosolic GS1 activity and the decline of the chloroplastic GS2 activity. Our results suggest that both high irradiance and nitrogen deficiency accelerates senescence of the primary leaf, probably for maintaining the functionality of the young leaves, and that one of the reasons for this accelerated senescence may be the high cellular oxidation and oxidative damage caused by the earlier decline of the activity of the antioxidant enzymes in these plants (Pompelli et al., 2010).

5. Acknowledgment

This work was supported by Junta de Andalucía (grant P07-CVI-02627 and PAI group BIO-0159) and DGICYT (AGL2009-11290).

6. References

Agüera, E., Cabello, P., & de la Haba, P. (2010). Induction of leaf senescence by low nitrogen nutrition in sunflower (*Helianthus annuus* L.) plants. *Physiologia Plantarum*, Vol.138, pp. 256-267, ISSN 0031-9317

Balibrea-Lara, M.E., González-García, M.C., Fatima, T., Ehneß, R., Lee, T.K., Proels, R., Tanner, W., & Roitsch, T. (2004). Extracellular invertase is an essential component of cytokinin-mediated delay of senescence. *Plant Cell*, Vol.16, pp. 1276-1287, ISSN 1040-4651

Berger, S., Weichert, H., Porzel, A., Wasternack, C., Kühn, H., & Feussner, I. (2001). Enzymatic and non-enzymatic lipid peroxidation in leaf development. *Biochimica et Biophysica Acta*, Vol.1533, pp. 266-276, ISSN 0304-4165

Bleecker, A.P., & Patterson, S.E. (1997). Last exit: Senescence, abscission, and meristem arrest in *Arabidopsis*. *Plant Cell*, Vol.9, pp. 1169-1179, ISSN 1040-4651

Buchanan-Wollaston, V. (1997). The molecular biology of leaf senescence. *Journal of Experimental Botany*, Vol.48, pp. 181-199, ISSN 0022-0957

Buchanan-Wollaston, V., Earl, S., Harrison, E., Mathas, E., Navabpour, S., Page, T., & Pink, D. (2003a). The molecular analysis of leaf senescence- a genomics approach. *Plant Biotechnology Journal*, Vol.1, pp. 3-22, ISSN 1467-7644

Buchanan-Wollaston, V., Wellesbourne, H.R.I., & Warwick, U.K. (2003b). Senescence, leaves, In: *Encyclopedia of Applied Plant Sciences*, Elsevier Academic Press, 808-816

Brugière, N., Dubois, F., Masclaux, C., Sangwan, R.S., & Hirel, B. (2000) Immunolocalization of glutamine synthetase in senescing tobacco (*Nicotiana tabacum* L.) leaves suggests that ammonia assimilation is progressively shifted to the mesophyll cytosol. *Planta*, Vol. 211, pp. 519-527, ISSN 0032-0935

Cabello, P., de la Haba, P., & Maldonado, J.M. (1991). Isoforms of glutamine synthetase in cotyledons, leaves and roots of sunflower plants. *Journal of Plant Physiology*, Vol.137, pp. 378-380, ISSN 0176-1617

Cabello, P., Agüera, E., & de la Haba, P. (2006). Metabolic changes during natural ageing in sunflower (*Helianthus annuus*) leaves: expression and activity of glutamine synthetase isoforms are regulated differently during senescence. *Physiologia Plantarum*, Vol.128, pp. 175-185, ISSN 0031-9317

Couée, I., Sulmon, C., Gouesbet, G., & El-Amrani, A. (2006). Involvement of soluble sugars in reactive oxygen species balance and responses to oxidative stress in plants. *Journal of Experimental Botany*, Vol.3, pp. 449-459, ISSN 0022-0957

Dertinger, U., Schaz, U., & Schulze, E.D. (2003). Age-dependence of the antioxidative system in tobacco with enhanced glutathione reductase activity or senescence-induced production of cytokinins. *Physiologia Plantarum*, Vol.119, pp. 19-29, ISSN 0031-9317

Dhindsa, R.A., Plumb-Dhindsa, P., & Thorpe, T.A. (1981). Leaf senescence: correlated with increased permeability and lipid peroxidation, and decreased levels of superoxide dismutase and catalase. *Journal of Experimental Botany*, Vol.126, pp. 93-101, ISSN 0022-0957

Diaz, C., Purdy, S., Christ, A., Morot-Gaudry, J.F., Wingler, A., & Masclaux-Daubresse, C. (2005). Characterization of markers to determine the extent and variability of leaf senescence in Arabidopsis. A metabolic profiling approach. *Plant Physiology*, Vol.138, pp. 898-908, ISSN 0032-0889

Diaz, C., Lemaître, T., Christ, A., Azzopardi, M., Kato, Y., Sato, F., Morot-Gaudry, J.F., Le Dily, F., & Masclaux-Daubresse, C. (2008). Nitrogen recycling and remobilization are differentially controlled by leaf senescence and development stage in *Arabidopsis* under low nitrogen nutrition. *Plant Physiology*, Vol.147, pp. 1437-1449, ISSN 0032-0889

Gan, S., & Amasino, R.M. (1997). Making sense of senescence. *Plant Physiology*, Vol.113, pp. 313-319, ISSN 0032-0889

Gepstein, S. (1988). Photosynthesis, In: *Senescence and Aging in Plants*, L.D. Noodén & A.C. Leopold, (Eds.), 85-109, Academic Press Publishers, San Diego, USA

Guiboileau, A., Sormani, R., Meyer, C., & Masclaux-Daubresse, C. (2010). Senescence and death of plant organs: nutrient recycling and developmental regulation. *Comptes Rendus Biologies*, Vol.333, pp. 382-391 ISSN 1631-0691

Guo, I., & Gan, S. (2005). Leaf senescence: signals, execution, and regulation. *Current Topics in Developmental Biology*, Vol.71, pp. 83-112, ISSN 0070-2153

Hernández, J.A., Escobar, C., Creissen, G., & Mullineaux, P.M. (2004). Role of hydrogen peroxide and the redox state of ascorbate in the induction of antioxidant enzymes in pea leaves under excess light stress. *Functional Plant Biology*, Vol.31, pp. 359-368, ISSN 1445-4408

Hernández, J.A., Escobar, C., Creissen, G., & Mullineaux, P.M. (2006). Antioxidant enzyme in pea plants under high irradiance. *Biologia Plantarum*, Vol.50, pp. 395-399, ISSN 0006-3134

Hidema, J., Makino, A., Mae, T., & Ojima, T. (1991). Photosynthetic characteristics of rice leaves aged under different irradiances from full expansion through senescence. *Plant Physiology*, Vol.97, pp.1287-1293, ISSN 0032-0889

Himelblau, E., & Amasino, R.M. (2001). Nutrients mobilized from leaves of *Arabidopsis thaliana* during senescence. *Journal of Plant Physiology*, Vol.158, pp. 1317-1323, ISSN 0176-1617

Hörtensteiner, S. (2006). Chlorophyll degradation during senescence. *Annual Review of Plant Biology*, Vol.57, pp. 55-77, ISSN 1543-5008

Hörtensteiner, S., & Feller, U. (2002). Nitrogen metabolism and remobilization during senescence. *Journal of Experimental Botany*, Vol.53, pp. 927-937, ISSN 0022-0957

Ishida, H., Nishimori, Y., Sugisawa, M., Makino, A., & Mae, T. (1997). The large subunit of ribulose-1,5-bisphosphate carboxilase/oxygenase is fragmented into 37 kDa and 16 kDa polypeptides by active oxygen in the lysates of chloroplasts from primary leaves of wheat. *Plant and Cell Physiology*, Vol. 38, pp. 471-479, ISSN 0032-0781

Ishida, H., Anzawa, D., Kokubun, N., Makino, A., & Mae, T. (2002). Direct evidence for non-enzymatic fragmentation of chloroplastic glutamine synthetase by a reactive oxygen species. *Plant Cell and Environment*, Vol.25, pp. 625-631, ISSN 0140-7791

Jing, H.C., Hebeler, R., Oeljeklaus, S., Sitek, B., Stühler, K., Meyer, H.E., Sturre, M.J., Hille, J., Warscheid, B., & Dijkwell, P.P. (2008). Early leaf senescence is associated with an altered cellular redox balance in *Arabidopsis* cpr5/old1 mutants. *Plant Biology*, Vol.1, pp. 85-98, ISSN 1435-8603

Jongebloed, U., Szederkényi, J., Hartig, K., Schobert, C., & Komor, E. (2004). Sequence of morphological and physiological events during natural ageing and senescence of a castor bean leaf: sieve tube occlusion and carbohydrate back-up precede chlorophyll degradation. *Physiologia Plantarum*, Vol.120, pp. 338-346, ISSN 0031-9317

Kamachi, K., Yamaya, T., Mae, T., & Ojiva, K. (1991). A role for glutamine synthetase in the remobilization of leaf nitrogen during natural senescence in rice leaves. *Plant Physiology*, Vol.96, pp. 411-417, ISSN 0032-0889

Kamachi, K., Yamaya, T., Hayakawa, T., Mae, T., & Ojima, K. (1992). Changes in cytosolic glutamine synthetase polypeptide and its mRNA in a leaf blade of rice plants during natural senescence. *Plant Physiology*, Vol.98, pp. 1323-1329, ISSN 0032-0889

Kato, Y., Murakami, S., Yamamoto, Y., Chatani, H., Kondo, Y., Nakano, T., Yokota, A., & Sato, F. (2004). The DNA-binding protease, CND41, and the degradation of ribulose-1,5-bisphosphate carboxylase/oxygenase in senescent leaves of tobacco. *Planta*, Vol.220, pp. 97-104, ISSN 0032-0935

Kato, Y., Yamamoto, Y., Murakami, S., Sato, F. (2005). Post-translational regulation of CND41 protease activity in senescent tobacco leaves. *Planta*, Vol.222, pp. 643-651, ISSN 0032-0935

Kukavica, B., & Veljovic-Jovanovic, S. (2004). Senescence-related changes in the antioxidant status of ginkgo and birch leaves during autumn yellowing. *Physiologia Plantarum*, Vol.122, pp. 321-327, ISSN 0031-9317

Lim, P.O., Woo, H.R., & Nam, H.G. (2003). Molecular genetics of leaf senescence in *Arabidopsis*. *Trends in Plant Science*, Vol.8, pp. 272-278, ISSN 1360-1385

Lim, P.O., Kim, H.J., & Nam, H.G. (2007). Leaf senescence. Annual Review of Plant Biology, Vol.58, pp. 115-136, ISSN 1543-5008

Lin, J.F., & Wu, S.H. (2004). Molecular events in senescing *Arabidopsis* leaves. *Plant Journal*, Vol.39, pp. 612-628, ISSN 0960-7412

Martínez, D.E., Costa, M.L., & Guiamet, J.J. (2008). Senescence-associated degradation of chloroplast proteins inside and outside the organelle. *Plant Biology*, Vol.1, pp.15-22, ISSN 1435-8603

Masclaux, C., Valadier, M.H., Brugière, N., Morot-Gaudry, J.F., & Hirel, B. (2000). Characterization of the sink/source transition in tobacco (*Nicotiana tabacum* L.) shoots in relation to nitrogen management and leaf senescence. *Planta*, Vol.211, pp. 510-518, ISSN 0032-0935

Masclaux-Daubresse, C., Carrayol, E., & Valadier, M.H. (2005). The two nitrogen mobilization- and senescence-associated GS1 and GDH genes are controlled by C and N metabolites. *Planta*, Vol.221, pp. 580–588, ISSN 0032-0935

Masclaux-Daubresse, C., Purdy, S., Lemaître, T., Pourtau, N., Naconnat, L., Renou, J.P., & Wingler, A. (2007). Genetic variations suggest interaction between cold acclimation and metabolic regulation of leaf senescence. *Plant Physiology*, Vol.143, pp. 434-446, ISSN 0032-0889

Miller, A., Schlagnhaufer, C., Spalding, M., & Rodermel, S. (2000). Carbohydrate regulation of leaf development: prolongation of leaf senescence in Rubisco antisense mutants of tobacco. *Photosynthesis Research*, Vol.63, pp. 1-8, ISSN 0166-8595

Moore, B., Zhou, L., Rollanf, F., Hall, Q., Cheng, W.H., Líu, Y.X., Hwang, I., Jones, T., & Sheen, J. (2003). Role of the *Arabidopsis* glucose sensor HXK1 in nutrient, light and hormonal signalling. *Science*, Vol.300, pp. 332-336, ISSN 0036-8075

Morcuende, R., Krapp, A., Hurry, V., & Stitt, M. (1998). Sucrose-feeding leads to increased rates of nitrate assimilation, increased rates of α-oxoglutarate synthesis, and increased synthesis of a wide spectrum of amino acids in tobacco leaves. *Planta*, Vol.206, pp. 394-409, ISSN 0032-0935

Munné-Bosch, S., & Alegre, L. (2002). Plant aging increases oxidative stress in chloroplasts. *Planta*, Vol.214, pp. 608-615, ISSN 0032-0935

Nam, H.G. (1997). The molecular genetic analysis of leaf senescence. *Current Opinion in Biotechnology*, Vol.8, pp. 200-207, ISSN 0958-1669

Noodén, L.D., Guiamét, J.J., & John, I. (1997). Senescence mechanisms. *Physiologia Plantarum*, Vol.101, pp. 746-753, ISSN 0031-9317

Olsson, M. (1995). Alteration in lipid composition and antioxidative protection during senescence in drought stressed plants and non-drought stressed plants of *Pisum sativum*. *Plant Physiology and Biochemistry*, Vol.33, pp. 547-553, ISSN 0981-9428

Ortega, J.L., Roche, D., & Sengupta-Gopalan, C. (1999). Oxidative turnover of soybean root glutamine synthetase. In vitro and in vivo studies. *Plant Physiology*, Vol.119, pp. 1483-1495, ISSN 0032-0889

Otegui, M., Noh, Y.S., Martínez, D.E., Vila-Petroff, M., Staehelin, A., Amasino, R., & Guiamet, J.J. (2005). Senescence-associated vacuoles with intense proteolytic activity develop in senescing leaves of *Arabidopsis* and soybean. *Plant Journal*, Vol.41, pp. 831-844, ISSN 0960-7412

Ougham, H., Hörtensteiner, S., Armstead, I., Donnison, I., King, I., Thomas, H., & Mur, L. (2008). The control of chlorophyll catabolism and the status of yellowing as a biomarker of leaf senescence. *Plant Biology*, Vol.10, pp. 4-14, ISSN 1435-8603

Palatnik, J.F., Carrillo, N., & Valle, E.M. (1999). The role of photosynthetic electron transport in the oxidative degradation of chloroplastic glutamine synthetase. *Plant Physiology*, Vol.121, pp. 471-478, ISSN 0032-0889

Parrott, D., Yang, L., Shama, L., & Fischer, A.M. (2005). Senecence is accelerated, and several proteases are induced by carbon "feast" conditions in barley (*Hordeum vulgare* L.) leaves. *Planta*, Vol.222, pp. 989-1000, ISSN 0032-0935

Pérez-Rodríguez, J., & Valpuesta, V. (1996). Expression of glutamine synthetase genes during natural senescence of tomato leaves. *Physiologia Plantarum*, Vol.97, pp. 576-582, ISSN 0031-9317

Pompelli, M.F., Martins, S., Antunes, W.C., Chaves, A., & DaMatta, F.M. (2010). Photosynthesis and photoprotection in coffee leaves is affected by nitrogen and light availabilities in winter conditions. *Journal of Plant Physiology*, Vol.167, pp 1052-1060, ISSN 0176-1617

Pourtau, N., Jennings, R., Pelzer, E., Pallas, J., & Wingler, A. (2006). Effect of sugar-induced senescence on gene expression and implications for the regulation of senescence in *Arabidopsis*. *Planta*, Vol.224, pp. 556-568, ISSN 0032-0935

Prochazkova, D., Sairam, R.K., Srivastava, G.C., & Singh, D.V. (2001). Oxidative stress and antioxidant activity as the basis of senescence in maize leaves. *Plant Science*, Vol.161, pp. 765-771, ISSN 0168-9452

Procházkova, D., & Wilhelmova, N. (2007). Leaf senescence and activities of the antioxidant enzymes. *Biologia Plantarum*, Vol.51, pp. 401-406, ISNN 0006-3134

Quirino, B.F., Noh, Y.S., Himelblau, E., & Amasino, R.M. (2000). Molecular aspects of leaf senescence. *Trends in Plant Science*, Vol.5, pp. 278-282, ISSN 1360-1385

Roulin, S., & Feller, U. (1998). Dithiothreitol triggers photo-oxidative stress and fragmentation of the large subunit of ribulose-1,5-bisphosphate carboxylase/oxygenase in intact pea chloroplasts. *Plant Physiology and Biochemistry*, Vol.36, pp. 849-856, ISSN 0981-9428

Schlüter, T., Leide, J., & Conrad, K. (2011). Light promotes an increase of cytokinin oxidase/dehydrogenase activity during senescence of barley leaf segments. *Journal of Plant Physiology*, Vol.168, pp. 694-698, ISSN 0176-1617

Špundová, M., Popelková, H., Ilík, P., Skotnica, J., Novotný, R., & Nauš, J. (2003). Ultra-structural and functional changes in the chloroplasts of detached barley leaves senescing under dark and light conditions. *Journal of Plant Physiology*, Vol.160, pp. 1051-1058, ISSN 0176-1617

Špundová, M., Sloukova, K., Hunková, M., & Nauš, J. (2005). Plant shading increases lipid peroxidation and intensifies senescence-induced changes in photosynthesis and activities of ascorbate peroxidase and glutathione reductase in wheat. *Photosynthetica*, Vol.43, pp. 403-409, ISSN 0300-3604

Srivalli, S., & Khanna-Chopra, R. (2009). Delayed wheat flat leaf senescence due to removal of spikelets is associated with increased activities of leaf antioxidant enzymes, reduced glutathione/oxidized glutathione ratio and oxidative damage to mitochondrial proteins. *Plant Physiology and Biochemistry*, Vol.47, pp. 663-670, ISSN 0981-9428

Streit, L., & Feller, U. (1983). Changing activities and different resistance to proteolytic activity of two forms of glutamine synthetase in wheat leaves during senescence. *Physiologie Végétale*, Vol.21, pp. 103-108, ISSN 0031-9368

Suzuki, Y., & Shioi, Y. (2004). Changes in chlorophyll and carotenoid contents in radish (*Raphanus sativus*) cotyledons show different time courses during senescence. *Physiologia Plantarum*, Vol.122, pp. 291-296, ISSN 0031-9317

Taiz, L., & Zeiger, E. (2010). *Plant Physiology* (5th edition). Sinauer Associates Inc., ISBN 978-0-87893-866-7, Sunderland, Massachusetts USA

Tercé-Laforgue, T., Mäck, G., & Hirel, B. (2004). New insights towards the function of glutamate dehydrogenase revealed during source-sink transition of tobacco (*Nicotiana tabacum*) plants grown under different nitrogen regimes. *Physiologia Plantarum*, Vol.120, pp. 220-228, ISSN 0031-9317

Van Doorn, V.G. (2008). Is the onset of senescence in leaf cells of intact plants due to low or high sugars level? *Journal of Experimental Botany*, Vol.59, pp. 1963-1972, ISSN 0022-0957

Van Lijsebettens, M., & Clarke, J. (1998). Leaf development in *Arabidopsis*. *Plant Physiology and Biochemistry*, Vol.36, pp. 47-60, ISSN 0981-9428

Vanacker, H., Sandalio, L.M., Jiménez, A., Palma, J.M., Corpas, F.J., Meseguer, V., Gómez, M., Sevilla, F., Leterrier, M., Foyer, C.H., & del Río, L.A. (2006). Roles for redox regulation in leaf senescence of pea plants grown on different sources of nitrogen nutrition. *Journal of Experimental Botany*, Vol.57, pp. 1735-1745, ISSN 0022-0957

Weaver, L.M., & Amasino, R.M. (2001). Senescence is induced in individually darkened *Arabidopsis* leaves, but inhibited in whole darkened plants. *Plant Physiology*, Vol.127, pp. 876-886, ISSN 0032-0889

Wiedemuth, K., Müller, J., Kahlau, A., Amme, S., Mock, H.-P., Grzam, A., Hell, R., Egle, K., Beschow, H., & Humbeck, K. (2005). Successive maduration and senescence of individual leaves during barley whole plant ontogeny reveals temporal and spatial regulation of photosynthetic function in conjunction with C and N metabolism. *Journal of Plant Physiology*, Vol.162, pp. 1226-1236, ISSN 0176-1617

Wingler, A., & Roitsch, T. (2008). Metabolic regulation of leaf senescence: interactions of sugar signalling with biotic and abiotic stress response. *Plant Biology*, Vol.10, pp. 50-62, ISSN 1435-8603

Wingler, A., Bownhill, E., & Pourtau, N. (2005). Mechanisms of light-dependent induction of cell death in tobacco plants with delayed senescence. *Journal of Experimental Botany*, Vol.56, pp. 2897-2905, ISSN 0022-0957

Wingler, A., Purdy, S., Maclean, J.A., & Pourtau, N. (2006). The role of sugars in integrating environmental signals during the regulation of leaf senescence. *Journal of Experimental Botany*, Vol.57, pp. 391-399, ISSN 0022-0957

Wingler, A., Maxclaux-Daubresse, C., & Fischer, A.M. (2009). Sugars, senescence, and ageing in plants and heterotrophic organisms. *Journal of Experimental Botany*, Vol.60, pp. 1063-1066, ISSN 0022-0957

Ye, Z., Rodríguez, R., Tran, A., Hoang, H., de los Santos, D., Brown, S., & Vellanoweth, R.L. (2000). The development transition to flowering repress ascorbate peroxidase activity and induces enzymatic lipid peroxidation in leaf tissue in *Arabidopsis thaliana*. *Plant Science*, Vol.158, pp. 115-127, ISSN 0168-9452

Yoo, S.D., Greer, D.H., Laing, W.A., & McManus, M.T. (2003). Changes in photosynthetic efficiency and carotenoid composition in leaves of white clover at different developmental stages. *Plant Physiology and Biochemistry*, Vol.4, pp. 887-893, ISSN 0981-9428

Yoshimura, K., Yabuta, Y., Ishikawa, T., & Shigeoka, S. (2000). Expression of spinach ascorbate peroxidase isoenzymes in response to oxidative stress. *Plant Physiology*, Vol.123, pp. 223-233, ISSN 0032-0889

Zapata, J.M., Guéra, A., Esteban-Carrasco, A., Martin, M., & Sabater, B. (2005). Chloroplasts regulate leaf senescence: delayed senescence in transgenic *ndh*F-defective tobacco. *Cell Death and Differentiation*, Vol.12, pp. 1277-1284, ISSN 1350-9047

Zimmermann, P., & Zentgraf, U. (2005). The correlation between oxidative stress and leaf senescence during plant development. *Cellular & Molecular Biology Letters*, Vol.10, pp. 515-534, ISSN 1425-8153

Zimmermann, P., Heinlein, C., Orendi, G., & Zentgraf, U. (2006). Senescence-specific regulation of catalases in *Arabidopsis thaliana* (L.) Heynh. *Plant Cell and Environment*, Vol.29, pp. 1049-1060, ISSN 0140-7791

5

Role of Intracellular Hydrogen Peroxide as Signalling Molecule for Plant Senescence

Ulrike Zentgraf, Petra Zimmermann and Anja Smykowski
ZMBP, University of Tübingen
Germany

1. Introduction

All aerobic organisms use molecular oxygen as terminal oxidant during respiration. Oxygen is neither very reactive nor harmful, but it has the potential to be only partially reduced, leading to the formation of very reactive and therefore toxic intermediates, like singlet oxygen ($^1O_2$), superoxide radical ($O_2^{\cdot-}$), hydroperoxylradical ($HO_2^{\cdot-}$), hydrogen peroxide (H_2O_2) and hydroxylradical ($\cdot OH$). These forms are called "reactive oxygen species" (ROS). All ROS are extremely reactive and may oxidize biological molecules, such as DNA, proteins and lipids. However, these reactive molecules are unavoidable by-products of an aerobic metabolism. It is known that reactive oxygen species may have a dual role in plant stress response (Dat et al. 2000). Whereas high concentrations of hydrogen peroxide are toxic for the cell, low concentrations may act as signal which triggers the plant response upon a variety of biotic and abiotic stresses (Dat et al., 2000; Grant & Loake, 2000). It has been known for many years that common signal transduction molecules like MAPKs and calmodulin play an important role in some of these ROS signal transduction pathways.

Mitochondria are an important origin of ROS. During respiration, the ubiquinone pool is the main source for superoxide production. The alternative oxidase (AOX) could be identified in plants and protists, e.g. Trypanosoma, fungi, like *Neurospora crassa* and *Hansenula anomala* and in green algae, e.g. in Chlamydomonas (McIntosh, 1994). It acts as a quinoloxidase by transferring electrons from the reduced ubiquinone directly to molecular oxygen forming water (Siedow & Moore, 1993). AOX mediates an energy-wasteful form of respiration, but its physiological significance is still a matter of intense debate (Rasmusson et al., 2009; Vanlerberghe et al., 2009; Millar et al., 2011). The plant alternative oxidases form homodimers (Moore et al., 2002) and are encoded by a small gene family. In *Arabidopsis thaliana* five genes are known, *AOX1a*, *AOX1b*, *AOX1c*, *AOX1d* and *AOX2*, each exhibiting organ specific expression (Saisho et al., 1997; https://www.genevestigator.com). Among these five *AOX* genes in *Arabidopsis thaliana*, *AOX1a* is the major isoform expressed in leaves (Clifton et al., 2006). One important function of the alternative oxidase is to prevent the formation of excess of reactive oxygen molecules (Maxwell et al., 1999). AOX ensures a low reduction status of the ubiquinone pool by oxidizing ubiquinol. Thus, the electron flow is guaranteed (Millenaar & Lambers, 2003). This reaction is necessary, if the cytochrome *c* dependent pathway is restricted by naturally occurring cyanide, NO, sulphide, high concentrations of CO_2, low temperatures or phosphorus deprivation (Millenaar & Lambers, 2003) as well as wounding, drought, osmotic stress, ripening and pathogen infection

(McIntosh, 1994; Moore et al., 2002). Photo-oxidative stress of chloroplasts is also involved in *AOX* up-regulation (Yoshida et al. 2008). Moreover, ascorbate biosynthesis in mitochondria is linked to the electron transport chain between complexes III and IV (Bartoli et al., 2000) and leaves of the *AOX*-overexpressing lines accumulate more ascorbic acid than wild-type leaves (Bartoli et al., 2006). A lack of AOX can lead to an up-regulation of transcripts of the antioxidant defense system at low temperature (Watanabe et al, 2008). Therefore, it is likely that AOX is an important component in antioxidant defense mechanisms.

In addition, it is proposed that AOX also has important functions outside the mitochondria (Arnholdt-Schmitt et al., 2006; Clifton et al., 2006; Van Aken et al., 2009). Furthermore, a beneficial role for AOX in illuminated leaves has been suggested and AOX-deficient *aox1a* mutant showed a lowered operating efficiency of photosystem II and an enhanced activity of cyclic electron transport around photosystem I (CET-PSI) at high irradiance (Yoshida et al., 2011). However, in most cases, transgenic plants with altered levels of AOX exhibited no obvious variation in plant growth phenotype (Vanlerberghe et al., 2009), implying that AOX does not severely affect photosynthetic carbon gain and biomass productivity. In addition, AOX also has an effect on the control of NO levels in plant cells (Wulff et al., 2009).

There is some evidence that alternative respiration is correlated with senescence and longevity. Aging potato slides showed a decline in the capacity of cytochrome *c* dependent respiration whereas the alternative respiration as well as the protein content of AOX increased (Hiser & McIntosh, 1990). Expression of *AOX1a* of Arabidopsis is highest in rosette leaves at the onset of senescence (https:/www.genevestigator.com). Interestingly, the inactivation of subunit V of the cytochrome *c* oxidase complex in the fungus *Podospora anserina* led to the exclusive use of the alternative respiration pathway and to a decline in ROS formation in these mutants. This inactivation of the cytochrome *c* oxidase resulted in an extraordinary longevity of this fungus (Dufour et al. 2000). There are several lines of evidence that beside mitochondria also chloroplasts and peroxisomes trigger leaf senescence. For peroxisomes a ROS-mediated function in leaf senescence has been described (del Río et al. 1998). Tobacco deficient in the thylacoid Ndh complex showed a delay in leaf senescence. It was discussed that the senescence delay was achieved by lower ROS production (Zapater et al., 2005).

In different Arabidopsis mutants a tight correlation between extended longevity and tolerance against oxidative stress has been observed (Kurepa et al., 1998). The most extended longevity mutant of this collection which also showed the highest tolerance against paraquat treatment was *gigantea3*. GIGANTEA acts in blue light signaling and has biochemically separable roles in circadian clock and flowering time regulation (Martin-Tryon et al., 2007). However, the link between this nuclear localized protein and resistance to oxidative stress is still unclear. CATALASE2 (CAT2) and CATALASE3 (CAT3) enzymes, which are expressed under the control of the circadian clock, might be good candidates. They exhibit a higher activity in the *gigantea3* mutant which might be responsible for the elevated oxidative stress tolerance (Zentgraf & Hemleben, 2007). In contrast, the delayed leaf senescence mutants of Arabidopsis *ore1*, *ore3*, and *ore9* also exhibit increased tolerance to various types of oxidative stress but the activities of antioxidant enzymes were similar or lower in the mutants, as compared to wild type providing evidence that oxidative stress tolerance is also genetically linked to control of leaf longevity in plants (Woo et al., 2004).

In addition, the expression of many SAGs is enhanced by increased levels of reactive oxygen species (Miller et al., 1999; Navabpour et al., 2003) indicating that elevated levels of ROS might be used as a signal to promote senescence. In Arabidopsis the coordinate regulation of the hydrogen peroxide scavenging enzymes catalase (CAT) and ascorbate peroxidase APX) leads to a defined increase of hydrogen peroxide content during bolting time (Ye et al., 2000; Zimmermann et al., 2006). Removing the bolt and thereby delaying the decrease in APX activity led to a delay in chlorophyll degradation and senescence (Ye et al., 2000). Since APX enzyme activity appears to be regulated on the posttranscriptional level (Panchuk et l., 2005; Zimmermann et al., 2006) and appears to be inhibited by hydrogen peroxide itself in this developmental stage, the initial event to create the hydrogen peroxide peak during bolting time at the onset of senescence is the transcriptional down-regulation of *CAT2*. The transcription factor responsible for this down-regulation was isolated by a yeast-one hybrid screen and turned out to be a member of the bZIP transcription factor family, namely GBF1. If *GBF1* is knocked out by a T-DNA insertion, the down-regulation of *CAT2* during bolting time is abolished, the hydrogen peroxide peak during bolting time disappears and senescence is delayed (Smykowski et al., 2010). This hydrogen peroxide peak is discussed to trigger senescence induction by activating the systemic expression of the senescence-related transcription factors e.g. *WRKY53* (Miao et al., 2004).

In order to understand the correlation between mitochondrial ROS production and senescence in *Arabidopsis thaliana*, we treated cell cultures and whole Arabidopsis plants with antimycin A, an inhibitor of cytochrom c oxidase, and measured hydrogen peroxide production and senescence parameters. In addition, two different genes encoding the peroxisomal enzyme catalase have been knocked-out and the single knock-out plants *cat2* and *cat3* have been crossed to produce double knock-out plants *cat2/3*. In these plants also the consequences on hydrogen peroxide levels and leaf senescence were analysed.

2. Results and discussion

2.1 Changes in mitochondrial hydrogen peroxide production

Dufour and others (2000) characterized an almost immortal mutant of the fungus *Podospora anserina* carrying a mutation in the gene encoding subunit V of the cytochrom c oxidase complex. These mutants exclusively used the alternative respiration pathway thus clearly leading to a lower content of reactive oxygen species than in normal growing fungi. In Arabidopsis resistance to oxidative stress and longevity are also tightly correlated (Kurepa et al., 1998; Woo et al., 2004). Therefore, we wanted to analyse Arabidopsis plants and cells with increased alternative respiration for mitochondrial ROS production and a delay in senescence.

2.1.1 Antimycin A treatment of cell cultures and whole plants

The alternative respiration in plants can be induced by application of antimycin A (Vanlerberghe & McIntosh, 1992), which was isolated from *Streptomyces* sp. and inhibits specifically the electron transport between cytochrome b and c_1. To investigate the influence of antimycin A on the production of ROS, we analysed antimycin A treated *Arabidopsis thaliana* cell cultures for their hydrogen peroxide contents. Two hours after treatment with 5 µM antimycin A or 0.02 % ethanol as control the H_2O_2 concentration slightly increased, whereas further incubation clearly lowered H_2O_2 content in antimycin A treated cells in comparison to control cells (Fig. 1 A). The transient increase in H_2O_2 levels might be a result of the inhibition of cytochrome c oxidase as it was already shown by Maxwell and coworkers (1999) for tobacco cells. Here an initial hydrogen peroxide production after antimycin A treatment could be localized almost exclusively to the mitochondria using laser scanning microscopy of H₂DCF-DA and mitotracker double-labelled cells. Dot blot analyses of 10 μg total RNA isolated from Arabidopsis culture cells and subsequent hybridization revealed that alternative oxidase 1a (AOX 1a) was induced by antimycin A as well as by hydrogen peroxide treatment (Fig. 1 B). This was also observed in tobacco cells, where antimycin A led to a more efficient alternative respiration capacity (Maxwell et al., 2002) and subsequently to a reduced mitochondrial ROS formation (Maxwell et al., 1999).

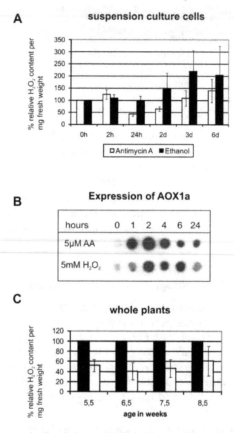

Fig. 1. *Short term treatment of cell cultures and whole plants with antimycin A.*
A) *Arabidopsis thaliana* cell cultures were treated with 5 μM antimycin A or 0.02 % ethanol as control and were analysed for their hydrogen peroxide content. The 0 h value was referred to as 100%. The error bars indicate the standard deviation of 4 independent experiments.
B) Hybridization of 10μg of total RNA isolated from antimycin A or H₂O₂ treated culture cells dotted on a nylon filter with an *AOX1a* specific probe. **C)** *Arabidopsis thaliana* plants of different developmental stages were watered with 10 ml of a 20 μmol antimycin A solution whereas control plants were treated with 0.8 % ethanol and were analysed for their hydrogen peroxide content after 24 hours. The values of ethanol treated plants were referred to as 100%. The error bars indicate the standard deviation of 4 independent experiments.

Since we were interested in analyzing the induction of senescence in whole Arabidopsis plants, we watered plants of different developmental stages with antimycin A and measured the hydrogen peroxide content 24 h after the treatment. In all developmental stages the hydrogen peroxide content was significantly lower in leaves of antimycin A treated plants (Fig. 1 C) indicating that in all developmental stages AOX and alternative respiration was induced to reduce mitochondrial ROS production.

2.1.2 Long term treatment of plants with antimycin A

In order to elucidate the long term effects of alternative respiration on plant development and senescence, soil grown Arabidopsis plants were watered over a time period of five weeks with 10 ml 20 μmol antimycin A solution every second day beginning with 5-week-old plants. Control plants were treated with 0.8 % ethanol in which antimycin A was dissolved. Since it was possible to reduce the hydrogen peroxide levels by the induction of the alternative pathways in all developmental stages, we assume that these plants grew under conditions favouring the alternative respiration from week 5 on. We have chosen this experimental design in order to guarantee that plant growth and development is not impaired in early stages by the lack of a functional cytochrome c pathway and the ATP it generates. Therefore, we did not use cytochrome c oxidase knock-out mutants, which appear to be impaired in growth and development from early on (data not shown).

The hydrogen peroxide content of antimycin A watered and control plants was measured weekly and the H_2O_2 level at the beginning of the experiment was set as 100 % (Fig. 2A). H_2O_2 concentrations of the control plants exhibit a peak in 7-week-old plants during the time of bolting and an increase in late stages of development as it was already shown before (Miao et al. 2004; Zimmermann et al. 2006).

In contrast, the antimycin A treated plants showed a slight decrease up to 8 weeks and recovered in older stages to the starting level (Fig. 2A). This coincides with the results of Dufour and coworkers (2000) for the fungus *Podospora anserine*, where long term activated alternative respiration led to lower hydrogen peroxide contents and strongly increased longevity. However, in Arabidopsis no obvious differences in the development and the progression of senescence could be detected phenotypically in antimycin A treated plants (Fig. 2B). In contrast, a transgenic Arabidopsis line overexpressing the senescence-associated transcription factor *WRKY53* exhibited an accelerated senescence phenotype (Fig. 2B; Miao et al., 2004). In accordance with the phenotype, chlorophyll and total protein content differed only slightly between antimycin A and ethanol treated plants, but were reduced earlier in 35S:*WRKY53* plants (Fig. 2C). Northern blot analyses revealed that the senescence-specific cystein protease gene *SAG12* was induced earlier and stronger in the antimycin A treated plants (Fig. 2D). This implies that even though less reactive oxygen species are produced in plants with favoured alternative respiration, development and senescence are not impaired or even slightly accelerated. Maxwell et al. (2002) presented evidence that, besides AOX, different senescence associated genes of tobacco (e.g. ACC, GST and Cystein protease precursor) can rapidly be induced by antimycin A treatment and this rapid induction can be prevented by ROS scavengers (Maxwell et al., 2002). In addition, overexpression of AOX in tobacco culture cells led to a decline in ROS concentration and a reduced expression of antioxidative enzymes, like superoxide dismutase or glutathione peroxidase (Maxwell et al., 1999).

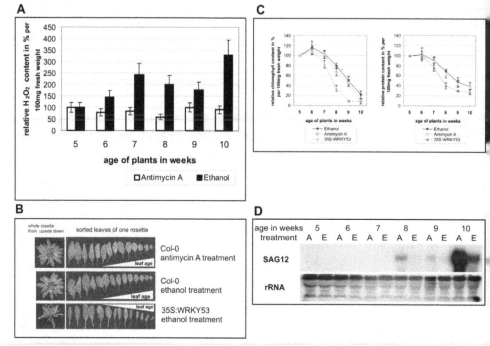

Fig. 2. *Long term treatment of whole plants with antimycin A.*
A) *Arabidopsis thaliana* plants were watered with 10 ml of a 20 µmol antimycin A solution every second day over a time period of five weeks beginning with 5-week-old plants. Control plants were treated with 0.8 % of ethanol. These plants were analysed for their hydrogen peroxide content every week. The values of the 5-week-old plants were referred to as 100%. The error bars indicate the standard deviation of 3 independent experiments. **B)** Phenotypic analyses of antimycin A or ethanol treated 8-week-old wildtype plants and ethanol treated transgenic *WRKY53* overexpressing line (35S:*WRKY53*). Whole plants are shown upside down to visualize older leaves of the rosette. In addition, the leaves were sorted according to their age using a specific colour code. **C)** Chlorophyll (left) and total protein (right) were measured in ethanol treated wildtype plants (Col-0), antimycin A treated wildtype plants and ethanol treated *WRKY53* overexpessing plants (35S:*WRKY53*). The values of 5-week-old plants were referred to as 100%. The error bars indicate the standard deviation of 3 independent experiments. **D)** Northern blot analyses of 15 µg of total RNA isolated from antimycin A (A) or ethanol treated (E) plants. The nylon filters were hybridized with a *SAG12* specific probe. Rehybridization with a 25S rRNA probe was used as loading control.

Transgenic tobacco culture cells carrying an antisense construct for *AOX* show an increased ROS formation and an elevated transcript abundance of catalase (Maxwell et al., 1999). In contrast, Umbach and coworkers (2005) observed that in *AOX* overexpressing or *AOX* antisense transgenic Arabidopsis lines transcript levels of the antioxidative enzymes MnSOD, organellar APX, cytosolic and organellar glutathione reductase and peroxiredoxins were not altered. This indicates that in Arabidopsis the lower production of ROS does not lead to compensatory reduction of oxidative stress enzymes. A senescence phenotype was also not observed in these lines.

2.1.3 Transgenic plants overexpressing *AOX1a*

The analysis of transgenic plants is helpful to gain more information about the function of a gene. For this reason, plants expressing the genes *AOX1a* under the constitutive 35S promoter were generated. This isoform was selected since it is strongly expressed in leaves. Plants of the T2 generation of these overexpressing lines were tested for *AOX1a* expression and three lines were obtained which overexpressed the transgene about 20-fold. The H_2O_2 content of these lines was analysed and a clear reduction in the hydrogen peroxide content in the transgenic lines could be measured (Fig.3B). Again, no obvious senescence phenotype could be detected (Fig. 3A). If at all, a slight acceleration of leaf senescence can be observed in the 35S:*AOX1a* lines. Fiorani et al. (2005) could observe a phenotype in 35S:*AOX1a* lines under low temperature conditions (12°C) with increased leaf area and larger rosettes. This could not be observed in our 35S:*AOX1a* lines under normal growth conditions. However, the cytochrom *c* dependent respiration is still functional in these plants probably masking the effect of increased levels of AOX.

Millenaar and Lambers (2003) describe that there is no clear positive correlation between the concentration of AOX protein and its activity *in vivo*, since an increase in protein formation does not change pyruvate concentration and the reduction state of ubiquinone, which are necessary for the activation of the AOX protein. For example, tobacco leaves infected with tobacco mosaic virus showed an increased AOX protein level but no change in activity of the alternative respiration (Lennon et al., 1997). In the transgenic plants overexpressing AOX, the capacity of the alternative respiration pathway appears to be elevated, but this does not necessarily reflect its activation. In the same line of evidence neither overexpression nor inactivation of AOX caused a change in ROS formation in the fungus *Podospora anserina* (Lorin et al., 2001). There was no effect on lifespan or senescence in the transgenic fungi either. However, in our transgenic lines the ROS production is clearly reduced indicating an activation of the alternative respiration pathway but nevertheless no effect on senescence could be observed.

Overexpression of *AOX* in tobacco culture cells leads to a decline in ROS concentration and a reduced expression of other antioxidative enzymes, like superoxide dismutase or glutathione peroxidase (Maxwell et al., 1999) whereas transgenic tobacco culture cells carrying an antisense construct for *AOX* show an increased ROS formation and an elevated transcript abundance of catalase (Maxwell et al., 1999). This would suggest that the plants would be either more sensitive or more resistant to oxidative stress. In contrast, in transgenic Arabidopsis lines either overexpressing *AOX* or an *AOX* antisense construct transcript levels of the antioxidative enzymes MnSOD, organellar APX, cytosolic and organellar glutathione reductase and peroxiredoxins were not altered (Umbach et al., 2005). In consistence with these findings, no altered resistance against oxidative stress could be observed in the transgenic 35S:*AOX1a* transgenic plants, which we germinated on MS plates and applied oxidative stress by spraying the seedlings with hydrogen peroxide (Fig. 3C).

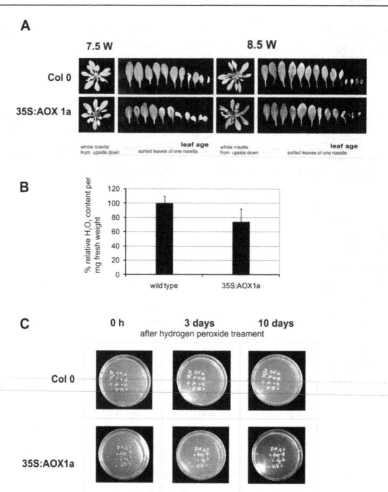

Fig. 3. *Transgenic plants overexperessing AOX1a*
A) Phenotypic analyses of wildtype (Col-0) and 35S:*AOX1a* transgenic plants. Whole plants
are shown upside down to visualize older leaves of the rosette. In addition, the leaves were
sorted according to their age using a specific colour code. **B)** 4-6 wildtype (Col-0) and
35S:*AOX1a* transgenic plants were pooled and analysed for their hydrogen peroxide
content. The values of the wild type plants were referred to as 100%. The error bars indicate
the standard deviation of 2 independently collected plant pools. **C)** Phenotypic analyses of
hydrogen peroxide treated seedlings of wild type (Col-0) and 35S:*AOX1a* transgenic plants.

2.1.4 Senescence-associated and circadian expression of *AOX1a*

The family of alternative oxidases comprises five genes with an organ specific expression
(Saisho et al., 1997; https://www.genevestigator.com). In general, *AOX* is expressed only at
a very low level under normal conditions. By using leaf material of plants of different age
harvested in the morning hours, a senescence dependent expression of *AOX1a* could be

bserved with the highest transcript abundance in old plants. In young, up to 7-week-old
lants, no expression could be detected by Northern blot analyses (Fig. 4A). This coincides
vith genevestigator data and with the *AOX* expression in different stages of the leaf
levelopment in potatoes, where an increase in AOX protein from young to mature leaves
ould be observed (Svensson & Rasmusson, 2001; https:/www.genevestigator.com).
'urthermore, there is a *de novo* synthesis of alternative oxidase in aging potato slides (Hiser
& McIntosh, 1990). Our *in silico* analysis of about 1500 bp upstream the coding region of the
AOX1a gene revealed, amongst others, several W-box core elements and one sequence for a
ircadian element. The W-boxes indicate a regulation by WRKY transcription factors which
ire involved in senescence or pathogen dependent regulation (Eulgem et al., 2000; Miao et
il., 2004) whereas the circadian element points out a clock dependent regulation. Based on
hese results, we used 8.5-week-old plants to harvest leaf material every three hours over 27
l. A circadian regulation of *AOX1a* could be detected with the maximum of expression in
he early morning hours with the beginning of illumination (Fig. 4B). This corresponds to
he expression of *AOX* in tobacco (Dutilleul et al., 2003).

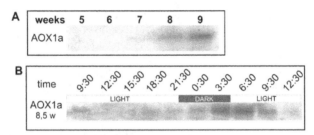

Fig. 4. *Senescence-associated expression of AOX1a*
Northern blot analyses of 15 μg of total RNA isolated from plants of **A)** 5-week-old to 9-
week-old plants and **B)** 8-week-old plants at different day times. The nylon filters were
hybridized with an *AOX1a* specific probe. Equal loading was controlled by Toluidin blue
staining of the membranes.

2.2 Changes in peroxisomal hydrogen peroxide production

Peroxisomes are organelles encircled by only a single membrane layer embedding an
extensive oxidative metabolism. These organelles are found in all eukaryotic organisms. In
plants, peroxisomes participate in many physiological processes like seed germination, leaf
senescence, fruit maturation, response to abiotic and biotic stress, photomorphogenesis,
biosynthesis of the plant hormones jasmonic acid and auxin, and in cell signaling by reactive
oxygen and nitrogen species. A specific feature of peroxisomes is their dynamic metabolism
meaning that the enzymatic constitution of peroxisomes is adjusted to the organism, cell or
tissue-type, and also to a variety of environmental conditions (Palma et al., 2009). One
important source for ROS formation, especially for H_2O_2, is photorespiration. During CO_2
fixation, ribulose-1,5-bisphosphate-carboxylase (RubisCO) can use CO_2 to carboxylate
ribulose-1,5-bisphosphate but also molecular oxygen to oxygenate ribulose-1,5-bisphosphate
forming glycolate. The glycolate is then transported from the chloroplasts into the
peroxisomes where it is oxidized generating H_2O_2 as a by-product. Peroxisomes and ROS
generated in these organelles were shown to play a central role in natural and dark induced
senescence in pea (del Rio et al., 1998) and appear to play an important role as a supplier of

signal molecules like NO· (nitric oxide), O_2^-, H_2O_2 and possibly S-nitrosoglutathione (del Rio et al., 1998; 2002; 2003). These signaling molecules can trigger specific gene expression by so far largely unknown signal transduction pathways (Corpas et al., 2001; 2004; del Rio et al. 2002). However, the concentration of these molecules is tightly regulated by a sensitive balance between production and decomposition by different specific scavenging systems. Catalases are the most abundant enzymes in peroxisomes and convert hydrogen peroxide into water and oxygen without the consumption of reducing equivalents. Besides catalases all enzymes of the antioxidant ascorbate-glutathione cycle, also called Foyer-Halliwell-Asada cycle, are present in peroxisomes to detoxify H_2O_2 through the oxidation of ascorbate and glutathione in an NADPH-dependent manner, thus complementing the action of catalase in peroxisomes. If the mitochondrial and the peroxisomal ascorbate-glutathione cycles are compared during progression of senescence, it can be speculated that peroxisomes may participate longer in the cellular oxidative mechanism of leaf senescence than mitochondria, since mitochondria appear to be affected by oxidative damage earlier than peroxisomes (Jiménez et al. , 1998; del Rio et al., 2003).

Catalases are tetrameric heme containing enzymes and are present in all aerobic organisms. Due to a very high apparent Michaelis constant catalases are not easily saturated with substrate and can act over a wide range of H_2O_2 concentrations maintaining a controlled intracellular H_2O_2 concentration. Whereas animals have only one form of catalase, plants have evolved small gene families encoding catalases. The plant catalases can be grouped into three classes depending on their expression and physiological parameters. In *Arabidopsis*, the small catalase gene family has been characterized to consist of three members, the class III catalase *CAT1*, class I catalase *CAT2* and class II catalase *CAT3*. All three Arabidopsis catalases show a senescence-specific alteration in expression and activity (Zimmermann et al., 2006). *CAT2* expression and activity is down-regulated at an early time point when plants are bolting. Subsequently, expression and activity of *CAT3* is up-regulated during progression of senescence. In contrast to *CAT2* expression, which is predominantly located in mesophyll cells, *CAT3* expression is mainly expressed in vascular tissue indicating that the vascular system appears to be protected against oxidative stress during senescence to guarantee the transport of nutrients and minerals out of the senescing tissue into developing parts of the plant like e.g. the seeds (Zimmermann et al., 2006). *CAT1* expression and activity is very low during plant development and only increases significantly during germination and in very late stages of senescence. Due to this expression pattern, its activity is discussed to be related to fatty acid degradation which takes place when peroxisomes are converted into glyoxisomes.

Especially the transcriptional down-regulation of *CAT2* appears to be involved in the regulation of the onset of senescence. This down-regulation is executed by the bZIP transcription factor GBF1. Insertion of a T-DNA into the *GBF1* gene revealed a loss of *CAT2* down-regulation and resulted in the loss of a hydrogen peroxide increase during bolting time. These *gbf1* mutant plants exhibit a delayed onset of senescence (Smykowski et al., 2010). Consequently, the idea suggests itself that *CAT2* knock-out plants also have a senescence phenotype. Taking into consideration that *CAT2* is expressed not only in leaves but also in roots, stems and flowers contributing substantially to the regulation of intracellular hydrogen peroxide contents and the protection of the cells against ROS in stress situations, the knock-out of this gene would have severe effects on the plants. The loss of such an important enzyme has to be compensated somehow during development but it would be expected that these knock-out plants are more sensitive against all stresses implying an increased ROS production and that they most likely show a senescence phenotype. The lack of peroxisomal catalase CTL-2 in *Caenorhabditis elegans* causes a progeric phenotype whereas the lack of the cytosolic catalase CTL-1 has no effect on nematode aging (Petriv & Rachubinski, 2004). In yeast, catalase T activity but not catalase A activity was necessary to assure longevity under repressing conditions on glucose media. However, under derepressing conditions, on ethanol media, both catalases were required for longevity assurance (Van Zandycke et al., 2002) indicating a correlation between CAT activity and longevity in animal systems.

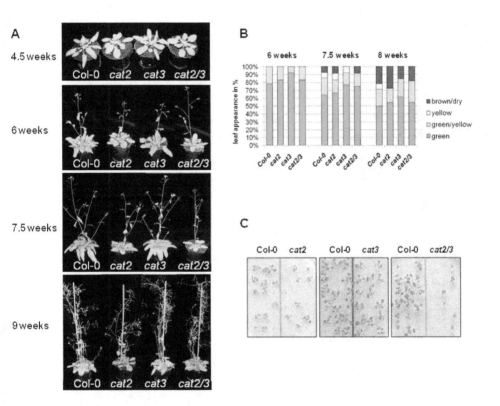

Fig. 5. *Phenotypic analyses of catalase mutants*
A) Plant development, **B)** Percantage of phenotypical appearance of the leaves of ten rosettes, **C)** Germination rate of wild type (Col-0), *cat2*, *cat3*, and *cat2/3* mutant plants.

Surprisingly, *cat2* knock-out plants appear to be more or less inconspicuous. They are slightly impaired in germination (Fig. 5C) but once germinated the plants developed relatively normaly (Queval et al., 2007; Fig. 5A). Photoperiod and CO_2 levels have a high impact on the phenotypic appearance of the plants and on the ascorbate and glutathione contents and their balances of the oxidized and reduced form, respectively. Under high CO_2

conditions no obvious phenotype could be observed whereas growth under ambient air, which favours photorespiration, led to a lower biomass production of the rosette and an altered leaf shape (Queval et al., 2007; Fig 5). We characterized SALK T-DNA insertion lines for *CAT2* and *CAT3* for homozygous insertion of the T-DNA and crossed the homozygous *cat2* and *cat3* mutants and selected the offsprings for a homozygous double knock-out line *cat2/3*. After separation of leaf protein extracts of these lines on native PAGEs, we could confirm that according to the gene knock-out the activity of the respective isoform disappeared (Fig. 6 A). When we analyzed plant development under long day conditions, leaf or plant senescence does not seem to be impaired (Fig. 5A, B); only leaf shape and biomass production were slightly altered in *cat2 and cat2/3* plants. However, the mutant plant populations did not senesce as homogenously as the wildtype populations. If the hydrogen peroxide content was measured, the profiles appeared to be not much different indicating that a very efficient compensation of the loss of CAT activity has been activated (Fig 6A).

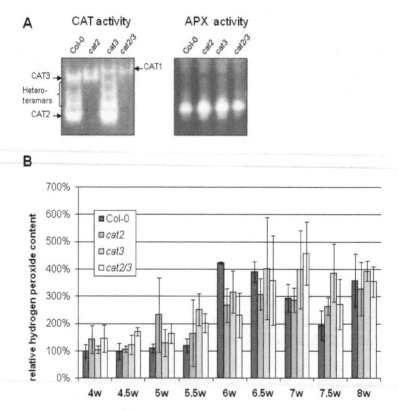

Fig. 6. *Physiological analyses of catalase mutants*
A) Catalase activity, **B)** Ascorbate peroxidase activity. **C)** Hydrogen peroxide content of wild type (Col-0), cat2, cat3, and cat2/3 mutant plants. The value of the 4–week-old wildtype plants was referred to as 100%. The error bars indicate the standard deviation of 4 independent experiments.

However, the hydrogen peroxide content between different leaves and plants varied remarkably so that the standard deviation was quite high in the measurements of the mutants. How the CAT2 or CAT3 activity losses were compensated is not yet clear but APX activity appeared to be not elevated (Fig 6B); in contrast, it appeared to be even slightly reduced in the double mutant. Hydrogen peroxide levels in these plants clearly indicated that the loss of the CAT activity must have been compensated. This is consistent with the finding of Rizhky and co-workers (2002), who claimed that there appears to be a sensitive balance between the antioxidant enzymes with compensating mechanisms, since they observed that double antisense plants for CAT or APX are more tolerant to oxidative stress than single antisense plants (Rizhsky et al., 2002). A slight activation of CAT1 can be observed in all our catalase mutants, especially in the *cat3* and *cat2/3* mutant plants. This is also indicated by the heterodimer formation between CAT2 and CAT1 in the *cat3* mutant. However, the activity of this isoform appears to be only low compared to the loss of catalase activity which would be present in wild type plants (Fig. 6B). Remarkably, glutathione levels are increased and shifted towards the more oxidized form in *cat2* plants under long day conditions (Queval et al., 2007). Taken together, the ROS levels appear to be very tightly regulated on many levels with the possibility of compensation if one detoxifying system fails.

3. Conclusion

Antimycin A treatment leads to the inhibition of the cytochrom *c* dependent electron transport lowering the production of hydrogen peroxide in mitochondria. Conversely, it is assumed that if stress occurs in a cellular compartment and increasing amounts of hydrogen peroxide are formed, these hydrogen peroxide molecules also can pass membranes and can be transported into the cytosol. This signal can then be transduced into the nucleus, where it induces the expression of many genes including *AOX*. As soon as the newly synthesized AOX protein is active, it minimizes the formation of ROS in the mitochondria by preventing the overreduction of the electron transport chain. Therefore, alternative oxidase might be regarded as mechanism to protect the plant from oxidative stress. Even though oxidative stress tolerance and longevity in Arabidopsis are tightly correlated (Kurepa et al. 1998, Woo et al., 2004) and hydrogen peroxide is discussed as signalling molecule to induce leaf senescence in Arabidopsis (Navabpour et al., 2003; Miao et al. 2004; Zimmermann et al., 2006), minimizing hydrogen peroxide production in the mitochondria by long-term antimycin A treatment did not delay senescence. In contrast, if down-regulation of *CAT2* expression and activity is abolished in *gbf1* mutants, the onset of senescence is delayed. On the other hand, if *CAT2* gene expression is prevented from early on in development in *cat2* T-DNA insertion lines, also no effect on senescence could be observed and hydrogen peroxide contents are not significantly altered. Therefore, we can assume that the intracellular origin but also the developmental time point of the hydrogen peroxide production might have an impact on its signalling function. In addition, the loss of one detoxifying system can be compensated by the cells and there seems to be a very sensitive balance between the different antioxidative protection systems. Remarkably, hydrogen peroxide plays a role in many different signal transduction pathways but how specificity is mediated is still an open question. Compartment-specific hydrogen peroxide fluorescent sensor molecules like roGFP or Hyper will help to clarify whether the intracellular origin of the hydrogen peroxide and changes during specific developmental time points might be important for its signalling function.

4. Experimental procedures

4.1 Plant material

Seeds from *Arabidopsis thaliana*, ecotype Columbia, were grown in a climatic chamber at 22°C under 16 h of illumination under low light conditions (60 μmol $s^{-1}m^{-2}$). Under these conditions plants developed flowers within 7 weeks, mature seeds could be harvested after 12 weeks. For long term treatment, plants were watered every second day with 5 ml of 40 μM antimycin A or 0.8 % ethanol as a control in addition to normal watering.

Suspension cells of *Arabidopsis thaliana*, ecotype Landsberg erecta, were grown under constant light on a rotary shaker (120 rpm) at 20°C and were subcultured every 7 days by 30-fold dilution in fresh growth medium (100 ml culture in 250 ml flasks). The Murashige and Skoog growth medium contains 3 % (w/v) sucrose, 0.5 mg/l α-naphthaleneacetic acid and 0.05 mg/l kinetin; pH was adjusted to 5.8 with KOH. Cell cultures with a density of about 100 mg/ml medium were treated with 5 μM antimycin A (Sigma) or 5 mM hydrogen peroxide.

The full length cDNA of AOX1a (At3g22370) was amplified by PCR form reverse transcripted poly A^+ RNA isolated from mature leaf material. The cDNA was cloned into the vector PY01 adjacent to a CaMV35S promoter. The construct was verified by sequencing. Arabidopsis transformation was performed by the vacuum infiltration procedure (Bechthold & Pelletier, 1998). The seeds of the transgenic plants were selected by spraying with 0.1% Basta. WRKY53 overexpressing plants were constructed as described before (Miao et al., 2004)

T-DNA insertion lines in *CAT2* (SALK_057998) and *CAT3* (SALK_092911) were obtained from the Nottingham Arabidopsis Stock Centre (NASC). Homozygous lines were characterized by PCR using gene specific and T-DNA left border primers (LBb1 5'GCGTGG ACC GCT TGC TGC AAC T 3'; CAT2-LP2 5' TCG CAT GAC TGT GGT TGG TTC 3'; CAT2-RP2 5' ACC ACC AAC TCT GGT GCT CCT 3'; CAT3-LP 5' CAC CTG AGT AAT CAA ATC TAC ACG 3'; CAT3-RP 5' TCA GGG ATC CTC TCT CTG GTG AA 3'). Homozygous plants were crossed and homozygous double knock-out lines were selected by PCR screening using the same primers. Knock-out was verified by native PAGE and subsequent CAT activity staining. Since *CAT2* and *CAT3* are under circadian regulation, leaves were always harvested 3 h after the beginning of illumination. Leaves were pooled in all experiments.

4.2 RNA isolation and Northern and dot blot analyses

Total RNA was isolated from leaves according to the protocol of PURESCRIPT RNA isolation kit (Gentra). Total RNA was either denatured 15 min at 55°C and spotted on nylon membranes or separated on MOPS-formaldehyde (6.2 %) agarose gels (1.5 %) and transferred to nylon membranes using 10 x SSC as transfer buffer. The membranes were hybridized at 65°C, washed twice at room temperature for 20 min with 2 x SSPE, 0.1 % SDS and once at 65°C for 30 min with 0.2 x SSPE, 0.1 % SDS. A fragment of the 5' UTR of the *AOX1a* gene or of the 3´UTR of the *SAG12* gene (At5g45890) was used as radioactive labeled hybridization probe.

4.3 Measurement and detection of hydrogen peroxide

Hydrogen peroxide was measured according to the method described by Kuźniak and others (1999). Ten leaf discs (diameter 1 cm) or pelleted suspension cells (approx. 100 mg) vere incubated for 2 h in 2 ml reagent mixture containing 50 mM potassium phosphate uffer pH 7.0, 0.05 % guaiacol (Sigma) and horseradish peroxidase (2.5 u/ml, Serva) at room emperature in the dark. Four moles of hydrogen peroxide are required to form 1 mole of etraguaiacol, which has an extinction coefficient of ε = 26.6 cm^{-1}mM^{-1} at 470 nm. The bsorbance in the reaction mixture was measured immediately at 470 nm.

4.4 Chlorophyll and total protein content

Leaf discs were homogenized in 0.2 ml 25 mM potassium phosphate buffer, pH 7.0, containing 2 mM EDTA. Subsequently, 0.8 ml acetone was added, and the samples were shaken vigorously for 1 h at room temperature. After centrifugation at 14000 g for 30 min at room temperature, the total chlorophyll content of the supernatant was measured and calculated following the method described by Arnon (1949). To determine total protein content, leaf discs were ground in 100 mM potassium phosphate buffer, pH 7.0, containing 1 mM EDTA at 4°C. After centrifugation at 14000 g for 30 min at 4°C, the supernatant was directly used for protein quantification according to the method of Bradford (1976) using BSA as standard.

4.5 CAT and APX acitivities

For analyses of APX isozymes, crude protein extracts were separated on 10% native polyacrylamide gels (0.375 M Tris-HCl, pH 8.8, as gel buffer) with a 5% stacking gel (0.125 M Tris-HCl, pH 6.8, as gel buffer) for 16 h (120V) at 4°C using 2 mM ascorbate, 250 mM glycine, and 25 mM Tris-HCl, pH 8.3, as electrophoresis buffer. After electrophoresis, the gels were soaked in 50 mM potassium phosphate buffer, pH 7.0, containing 2 mM ascorbate for 10 min (3x) and, subsequently, in 50 mM potassium phosphate buffer, pH 7.0, containing 4 mM ascorbate, and 1 mM H$_2$O$_2$ for 20 min. After rinsing in water, the gels were stained in 50 mM potassium phosphate buffer, pH 7.8, containing 14 mM TEMED (N,N,N´,N´-tetramethylethylenediamine) and 2.45 mM NBT (nitro blue tetrazolium) for 10-30 min. For the analyses of CAT isozymes the protein extracts were separated on 7.5% native polyacrylamide gels (0.375 M Tris-HCl, pH 8.8, as gel buffer) with a 3.5% stacking gel (0.125 M Tris-HCl, pH 6.8, as gel buffer) for 16 h (70-80V) at 18°C using 250 mM glycine and 25 mM Tris-HCl, pH 8.3, as electrophoresis buffer. Subsequently, the gels were stained for the activity of catalases as follows: The gels were soaked in 0.01% of hydrogen peroxide solution for 5 min, washed twice in water and incubated for 5 min in 1% FeCl$_3$ and 1% K$_3$[Fe(CN)$_6$]. After staining, the gels were washed once more in water.

5. Acknowledgment

We thank the Nottingham Arabidopsis Stock Centre (NASC) for providing seeds of the T-DNA insertion lines for *CAT2* (SALK_057998) and *CAT3* (SALK_092911). This work was supported by the Deutsche Forschungsgemeinschaft (SFB 446).

6. References

Arnholdt-Schmitt, B.; Costa, J.H. & de Melo, D.F. (2006). AOX- a functional marker for efficient cell reprogramming under stress? *Trends Plant Sciences* 11(6), pp. 281-287, ISSN 1360-1385

Arnon, D.I. (1949). Copper enzymes in isolated chloroplast. Polyphenoloxidase in *Beta vulgaris*. *Plant Physiology* 24, pp. 1-15, ISSN 0032-0889, online ISSN 1532-2548

Bartoli ,C.G.; Pastori ,G.M. & Foyer, C.H. (2000). Ascorbate biosynthesis in mitochondria is linked to the electron transport chain between complexes III and IV. *Plant Physiology* 123, pp. 335–344, ISSN 0032-0889, online ISSN 1532-2548

Bartoli, C.G.; Yu, J.; Gómez, F.; Fernández, L.; McIntosh, L. & Foyer, C.H. (2006). Interrelationships between light and respiration in the control of ascorbic acid synthesis and accumulation in Arabidopsis thaliana leaves. *Journal of Experimental Botany* 57 (8), pp. 1621-1631, ISSN 0022-0957, online ISSN 1460-2431

Bechtold, N. & Pelletier, G. (1998). In planta Agrobacterium-mediated transformation of adult Arabidopsis thaliana plants by vacuum infiltration. *Methods in Molecular Biology* 82, pp. 259-266, ISSN 1064-3745

Bradford, M.M. (1976). A rapid and sensitive method for the quantification of microgram quantities of protein utilizing the principle of protein-dye binding. *Annals of Biochemistry* 72, pp. 248-254 , ISSN 0305-7364, online ISSN 1095-8290

Clifton, R.; Millar, A.H. & Whelan, J. (2006). Alternative oxidases in Arabidopsis: a comparative analysis of differential expression in the gene family provides new insights into function of non-phosphorylating bypasses. *Biochimica et Biophysica Acta (BBA)* 1757 (7), pp. 730-741, ISSN 0005-2728, online ISSN 1879-2650

Corpas, F.J.; Barroso, J.B. & del Río, L.A. (2001). Peroxisomes as a source of reactive oxygen species and nitric oxide signal molecules in plant cells. *Trends in Plant Science* 6 (4), pp. 145–150, ISSN 1360-1385

Dat, J.; Vandenabeele, S.; Vranová, E.; Van Montagu, M.; Inzé, D. & Van Breusegem, F. (2000). Dual action of the active oxygen species during plant stress responses. *Cellular and Molecular Life Sciences* 57 (5), pp. 779-795, ISSN 1420-682X, online ISSN 1420-9071

del Río, L.A.; Pastori, G.M.; Palma, J. M.; Sandalio, L.M.; Sevilla, F.; Corpas, F.J.; Jiménez A.; López-Huertas, E. & Hernández, J. A. (1998). The activated oxygen role of peroxisomes in senescence. *Plant Physiology* 116 (4), pp. 1195–1200, ISSN 0032-0889, online ISSN 1532-2548

del Río, L.A.; Corpas, F.J.; Sandalio, L.M. ; Palma, J.M. & Barroso J.B. (2003). Plant peroxisomes, reactive oxygen metabolism and nitric oxide. *IUBMB Life* 55(2), pp. 71-81, ISSN 1521-6543, online ISSN 1521-6551

del Río, L.A.; Corpas, F.J.; Sandalio, L.M.; Palma, J.M.; Gómez, M. & Barroso, B. (2002). Reactive oxygen species, antioxidant systems and nitric oxide in peroxisomes. *Journal of Experimental Botany* 53 (372), pp. 1255–1272, ISSN 0022-0957, online ISSN 1460-2431

Dufour, E.; Boulay, J.; Rincheval, V. & Sainsard-Chanet, A. (2000). A causal link betweeen respiration and senescence in *Podospora anserina*. *Proceedings of the National Academy of Sciences USA* 97 (8), pp. 4138-4143, ISSN 0027-8424, online ISSN 1091-6490

Dutilleul, C.; Garmier, M.; Noctor, G.; Mathieu, C.; Chétrit, P.; Foyer, C.H. & de Paepe, R. (2003). Leaf mitochondria modulate whole cell redox homeostasis, set antioxidant capacity, and determine stress resistance through altered signaling and diurnal regulation. *Plant Cell* 15 (5), pp. 1212-1226, ISSN 1040-4651, online ISSN 1532-298X

Eulgem, T.; Rushton, P. J.; Robatzek, S. & Somssich, I. E. (2000). The WRKY superfamily of plant transcription factors. *Trends in Plant Science* 5 (5), pp. 199-206, ISSN 1360-1385

Fiorani, F.; Umbach, A.L. & Siedow, J.N. (2005). The alternative oxidase of plant mitochondria is involved in the acclimation of shoot growth at low temperature. A study of Arabidopsis AOX1a transgenic plants. *Plant Physiology* 139 (4), pp. 1795-1805, ISSN 0032-0889, online ISSN 1532-2548

Grant, J.J. & Loake, G.J. (2000). Role of reactive oxygen intermediates and cognate redox signalling in disease resistance. *Plant Physiology* 124 (1), pp. 21-29, ISSN 0032-0889, online ISSN 1532-2548

Hiser, C. & McIntosh, L. (1990). Alternative oxidase of potato is an integral membrane protein synthesized de novo during aging of tuber slices. *Plant Physiology* 93 (1), pp. 312-318, ISSN 0032-0889, online ISSN 1532-2548

Jiménez, A.; Hernandez, J.A.; Pastori, G.; del Rio, L.A. & Sevilla, F. (1998). Role of the ascorbate-glutathione cycle of mitochondria and peroxisomes in the senescence of pea leaves. *Plant Physiology* 118 (4), pp. 1327-1335, ISSN 0032-0889, online 1532-2548

Kurepa, J.; Smalle, J.; Van Montagu, M. & Inzé, D. (1998). Oxidative stress tolerance and longevity in Arabidopsis: the late flowering mutant *gigantea* is tolerant to paraquat. *Plant Journal* 14 (6), pp. 759-764, Online ISSN: 1365-313X

Kuźniak, E.; Patykowski, J. & Urbanek, H. (1999). Involvement of the antioxidative system in tomato response to fusaric acid treatment. *Journal of Phytopathology* 147, pp. 385-390, ISSN 0931-1785, online 1439-0434

Lennon, A.M.; Neuenschwander, U.H.; Ribas-Carbo, M.; Giles, L.; Ryals, J.A. & Siedow, J. N. (1997). The effects of salicylic acid and tobacco mosaic virus infection on the alternative oxidase of tobacco. *Plant Physiology* 115 (2), pp. 783-791, ISSN 0032-0889, online ISSN 1532-2548

Lorin, S.; Dufour, E.; Boulay, J.; Begel, O.; Marsy, S. & Sainsard-Chanet, A. (2001). Overexpression of the alternative oxidase restores senescence and fertility in a long- lived respiration-deficient mutant of *Podospora anserina*. *Molecular Microbiology* 42 (5), pp. 1259-1267, ISSN 0950-382X, online ISSN 1365-2958

Martin-Tryon, E.L.; Kreps, J.A. & Harmer, S.L. (2007). GIGANTEA acts in blue light signaling and has biochemically separable roles in circadian clock and flowering time regulation. *Plant Physiology* 143 (1), pp. 473-486, ISSN 0032-0889, online 1532-2548

Maxwell, D.P.; Nickels, R. & McIntosh, L. (2002). Evidence of mitochondrial involvement in the transduction of signals required for the induction of genes associated with pathogen attack and senescence. *Plant Journal* 29 (3), pp. 269-279, online ISSN 1365-313X

Maxwell, D.P.; Wang, Y. & McIntosh, L. (1999). The alternative oxidase lowers mitochondrial reactive oxygen production in plant cells. *Proceedings of the National Academy of Sciences USA* 96 (14), pp. 8271-8276, ISSN 0027-8424, online ISSN 1091-6490

McIntosh, L. (1994). Molecular biology of the alternative oxidase. *Plant Physiology* 105 (3), pp. 781-786, ISSN 0032-0889, online ISSN 1532-2548

Miao, Y.; Laun, T.; Zimmermann, P. & Zentgraf, U. (2004). Targets of the WRKY53 transcription factor and its role during leaf senescence in *Arabidopsis thaliana*. *Plant Molecular Biology* 55 (6), pp. 853-867, ISSN 0167-4412, online ISSN: 1573-5028

Millar, A.H.; Whelan, J.; Soole, K.L. & Day, D.A. (2011). Organization and regulation of mitochondrial respiration in plants. *Annual Review of Plant Biology* 62, pp. 79–104, ISSN: 1543-5008

Millenaar, F.F. & Lambers, H. (2003). The Alternative Oxidase: *in vivo* Regulation and Function. *Plant Biology* 5 (1), pp. 2-15, ISSN 1435-8603, online ISSN 1438-8677

Miller, J.D.; Arteca, R. N. & Pell, E.J. (1999). Senescence-associated gene expression during ozone-induced leaf senescence in Arabidopsis. *Plant Physiology* 120 (4), pp. 1015-1024, ISSN 0032-0889, online 1532-2548

Moore, A.L.; Albury, M.S.; Crichton, P.G. & Affourtit, C. (2002). Function of the alternative oxidase: is it still a scavenger? *Trends in Plant Science* 7 (11), pp. 478-481, ISSN 1360-1385

Navabpour, S.; Morris, K.; Allen, R.; Harrison, E.; Mackerness, S.A.H. & Buchanan-Wollaston V. (2003). Expression of senescence-enhanced genes in response to oxidative stress. *Journal of Experimental Botany* 54 (391), pp. 2285-2292, ISSN 0022-0957, online 1460-2431

Palma, J.M.; Copras, F.J. & del Rio, L.A. (2009). Proteome of plant peroxisomes: new perspectives on the role of these organelles in cell biology. *Proteomics* 9 (9), pp. 2301-2312, ISSN 1615-9853, online ISSN 1615-9861

Panchuk, I. I.; Zentgraf, U. & Volkov, R. A .(2005). Expression of the *Apx* gene family during leaf senescence of *Arabidopsis thaliana*. *Planta* 222 (5), pp. 926-932, ISSN 0032-0935, online ISSN 1432-2048

Petriv, O.I. & Rachubinski, R.A. (2004). Lack of peroxisomal catalase causes a progeric phenotype in Caenorhabditis elegans. *Journal of Biological Chemistry* 279 (19), pp. 19996-20001, ISSN 0021-9258, online ISSN 1083-351X

Queval, G.; Issakidis-Bourguet, E.; Hoeberichts, F.A.; Vandorpe, M.; Gakière, B.; Vanacker, H.; Miginiac-Maslow, M.; Van Breusegem, F. & Noctor, G. (2007). Conditional oxidative stress responses in the Arabidopsis photorespiratory mutant cat2 demonstrate that redox state is a key modulator of daylength-dependent gene expression, and define photoperiod as a crucial factor in the regulation of H_2O_2-induced cell death. *Plant Journal* 52 (4), pp. 640-657, Online ISSN: 1365-313X

Rasmusson, A.G.; Fernie, A.R. & van Dongen, J.T. (2009). Alternative oxidase: a defence against metabolic fluctuations? *Physiologia Plantarum* 137 (4), pp. 371–382, ISSN 0031-9317, online ISSN 1399-3054

Rizhsky, L.; Hallak-Herr, E.; Van Breusegem, F.; Rachmilevitch, S.; Barr, J.E.; Rodermel, S.; Inzé, D. & Mittler, R. (2002). Double antisense plants lacking ascorbate peroxidase and catalase are less sensitive to oxidative stress than single antisense plants lacking ascorbate peroxidase or catalase. *Plant Journal* 32 (3), pp. 329-342, Online ISSN: 1365-313X

Saisho, D.; Nambara, E.; Naito, S.; Tsutsumi, N.; Hirai, A. & Nakazono, M. (1997). Characterization of the gene family for alternative oxidase from *Arabidopsis thaliana*. *Plant Molecular Biology* 35 (5), pp. 585-596, ISSN 0167-4412, online ISSN: 1573-5028

Siedow, J. N. & Moore, A. L. (1993). A kinetic model for the regulation of electron transfer through the cyanide-resistant pathway in plant mitochondria. *Biochimica et Biophysica Acta - Bioenergetics* 1142 (1-2), pp. 165-174, ISSN 0006-3002

Smykowski, A.; Zimmermann, P. & Zentgraf, U. (2010). G-Box binding Factor1 reduces CATALASE2 expression and regulates the onset of leaf senescence in Arabidopsis. *Plant Physiology* 153 (3), pp. 1321-1331, ISSN 0032-0889, online 1532-2548

Svensson, A. S. & Rasmusson, A. G. (2001). Light-dependent gene expression for proteins in the respiratory chain of potato leaves. *Plant Journal* 28 (1), pp. 73-82, Online ISSN 1365-313X

Umbach, A.L.; Fiorani, F. & Siedow, J.N. (2005). Characterization of transformed Arabidopsis with altered alternative oxidase levels and analysis of effects on reactive oxygen species in tissue. *Plant Physiology* 139 (4), pp. 1806-1820, ISSN 0032-0889, online ISSN 1532-2548

Van Aken, O.; Zhang ,B.; Carrie, C.; Uggalla ,V.; Paynter, E.; Giraud, E. & Whelan, J. (2009). Defining the mitochondrial stress response in Arabidopsis thaliana. *Molecular Plant* 2 (6), pp. 1310–1324, ISSN 1674-2052, online ISSN 1752-9867

Van Zandycke, S.M.; Sohier, P.J. & Smart, K.A. (2002). The impact of catalase expression on the replicative lifespan of Saccharomyces cerevisiae. *Mechanisms of Ageing and Development* 123 (4), pp. 365-373, ISSN 0047-637

Vanlerberghe, G. C. & McIntosh, L. (1992). Coordinate regulation of cytochrome and alternative pathway respiration in tobacco. *Plant Physiology* 100 (4), pp. 1846-1851, ISSN 0032-0889, online ISSN 1532-2548

Vanlerberghe, G.C.; Cvetkovska, M. & Wang, J. (2009). Is the maintenance of homeostatic mitochondrial signaling during stress a physiological role for alternative oxidase? *Physiologia Plantarum* 137 (4), pp. 392–406, ISSN 0031-9317, online ISSN 1399-3054

Watanabe, C.K.; Hachiya, T.; Terashima, I. & Noguchi, K. (2008). The lack of alternative oxidase at low temperature leads to a disruption of the balance in carbon and nitrogen metabolism, and to an up-regulation of anti-oxidant defense systems in Arabidopsis thaliana leaves. *Plant, Cell & Environment* 31 (8), pp. 1190–1202, ISSN 0140-7791, online ISSN1365-3040

Woo, H.R.; Kim, J.H.; Nam, H.G. & Lim, P.O. (2004). The delayed leaf senescence mutants of Arabidopsis, ore1, ore3, and ore9 are tolerant to oxidative stress. *Plant & Cell Physiology* 45 (7), pp. 923-932, ISSN 0140-7791, online ISSN1365-3040

Wulff, A.; Oliveira, H.C.; Saviani, E.E.& Salgado, I. (2009). Nitrite reduction and superoxide-dependent nitric oxide degradation by Arabidopsis mitochondria: influence of external NAD(P)H dehydrogenases and alternative oxidase in the control of nitric oxide levels. *Nitric Oxide* 21(2), pp. 132-139, ISSN 1089-8603

Ye, Z.Z.; Rodriguez, R.; Tran, A.; Hoang, H.; de los Santos, D.; Brown, S. & Vellanoweth, R.L. (2000). The developmental transition to flowering represses ascorbate peroxidase activity and induces enzymatic lipid peroxidation in leaf tissue in *Arabidopsis thaliana. Plant Science* 158 (1-2), pp. 115-127, ISSN 0168-9452

Yoshida, K.; Watanabe, C. Kato, Y.; Sakamoto, W. & Noguchi, K. (2008). Influence of chloroplastic photo-oxidative stress on mitochondrial alternative oxidase capacity and respiratory properties: a case study with Arabidopsis yellow variegated 2. *Plant & Cell Physiology* 49 (4), pp. 592-603, ISSN 0032-0781, online ISSN 1471-9053

Yoshida, K.; Watanabe, C.K.; Terashima, I. & Noguchi, K. (2011). Physiological impact of mitochondrial alternative oxidase on photosynthesis and growth in Arabidopsis thaliana. *Plant Cell & Environment*, Jun 24. doi: 10.1111/j.1365-3040.2011.02384.x, ISSN 0140-7791, online ISSN1365-3040

Zapata, J.M.; Guera, A.; Esteban-Carrasco, A.; Martin, M. & Sabater, B. (2005). Chloroplasts regulate leaf senescence: delayed senescence in transgenic ndhF-defective tobacco. *Cell Death and Differentiation* 12 (10), pp. 1277-1284, ISSN 1350-9047

Zentgraf; U. & Hemleben, V. (2007). Molecular cell biology: Are reactive oxygen species regulators of leaf senescence? In: Lüttge, U.; Beyschlag, W. & Murata, J. (Eds), Progress in Botany, Vol. 69, Springer, Berlin, Heidelberg, New York, pp. 117-138, ISBN 978-3-540-72953-2, ISSN 0340-4773

Zimmermann, P.; Orendi, G.; Heinlein, C. & Zentgraf, U. (2006). Senescence specific regulation of catalases in *Arabidopsis thaliana* (L.) Heynh. *Plant Cell & Environment* 29 (6), pp. 1049-1060, ISSN 0140-7791, online ISSN1365-3040

6

Functional Approaches to
Study Leaf Senescence in Sunflower

Paula Fernandez, Sebastián Moschen, Norma Paniego and Ruth A. Heinz
Biotechnology Institute - CICVyA- INTA Castelar
Argentina

1. Introduction

Senescence is an age-dependent process at the cellular, tissue, organ or organism level, leading to death at the end of the life span (Noodén 1988). Annual plants as grain and oil crops undergo a visual process towards the end of the reproductive stage that is accompanied by nutrient remobilization from leaf to developing seeds (Buchanan-Wollaston et al. 2003). The final stage of this process is leaf death but this is actively delayed until all nutrients have been removed and recycle through the process of developmental senescence. It have been documented that a delay in leaf senescence has an important impact on grain yield trough the maintenance of the photosynthetic leaf area during the reproductive stage in different crops (Ewing & Claverie 2000), including sunflower (Sadras et al. 2000; De la Vega et al. 2011). The potential yields of sunflower crop are far from the real ones in all Argentina productive regions. In Balcarce, for example, while the potential yields are estimated in 5,000 kg.ha-1, those obtained by the best producers only reach 3,000 kg.ha-1, and the average in the region ranges in 1,800 kg.ha-1 (Dosio & Aguirrezábal 2004). These differences could possibly be due to the inability of current hybrids to keep their green leaf area for long periods, which would allow greater use of the incident radiation during the grain filling period which plays an important role in determining the yield and oil concentration in sunflower (Dosio et al. 2000; Aguirrezábal et al. 2003).

Besides autonomous (internal) factors as age, reproductive stage and phytohormone levels, leaf senescence is hardly affected by environmental factors. Among these environmental factors, including extreme temperature, drought, shading, nutrient deficient and pathogen infection, the most limiting ones are water and nutrient availability (Gan & Amasino 1997; Sadras et al. 2000; Sadras et al. 2000; Dosio et al. 2003; Lim et al. 2003; Aguera et al. 2010).

During leaf senescence, critical and dramatic changes occurred in a highly regulated manner following a genetically programmed process of high complexity. Chlorophyll degradation, nutrient recycling and remobilization are preceded or paralleled by RNA and protein degradation. Even though leaf senescence has been widely recognized and accepted as a type of Programmed Cell Death (PCD) (Noodén & Leopold 1987), the onset and progression of senescence is accompanied by global changes in gene expression. Thus, deep extensive efforts have been achieved to reveal relevant molecular process by identifying and analysing

Senescence Associated Genes (SAGs) as prior tags to disclosure the core of this complex process (Kim et al. 2007). SAGs genes have been extensively studied in model plant species (Audic & Claverie 1997; Gepstein et al. 2003; Balazadeh et al. 2008; Hu et al. 2010) and in some agronomical relevant crops (Andersen et al. 2004; Conesa et al. 2005; Espinoza et al 2007). Yet, although senescence and ageing might be considered synonyms, a distinct reference was previously discussed because the former comprises all those degenerative changes and cellular degradation occurring with little or non-reference to death, whereas the latter is considered the final developmental stage culminating in death (Nooden & Leopold 1988; Shahri 2011). In the last year, considering this limitation, many efforts are being achieved to disclosure and obtain genomic information for this oil crop (Kane et al 2011) but complete sequence information are still no available.

Sunflower (*Helianthus annuus* L.) is one of the most relevant crops as source of edible oil and many efforts have been achieved to build up useful functional genomics tools for cultivated sunflower involving transcriptional and metabolic profiles (Fernandez et al. 2003; Cabello et al. 2006; Paniego et al. 2007; Fernandez et al. 2008; Peluffo et al. 2010). Although, molecular studies focused on the onset of the senescence process in sunflower leaf are scarce (Fernandez et al. 2003; Dezar et al. 2005; Manavella et al. 2006; Jobit et al. 2007; Paniego et al. 2007; Fernandez et al. 2008; Manavella et al. 2008; Peluffo et al. 2010; Fernandez et al. 2011). Thus, two different approaches are envisage for studying molecular events occurring during leaf senescence: the first strategy relays on the identification of sunflower SAGs based on a candidate gene approach while the second approach involves concerted gene expression studies based on high density oligonucleotide microarrays, whole transcriptome shotgun sequencing and microRNA detection by RNA-seq (Buermans et al. 2010; Dhahbi et al. 2011).

Leaf senescence is a complex and highly coordinated process (Noodén et al. 1997). Although symptoms have been explored, the involved processes and the mechanisms that control it have not been characterized yet (Buchanan-Wollaston et al. 2003). The distinctive symptom of leaf senescence is the breakdown of chloroplasts, therefore the decrease in chlorophyll content becomes a key indicator of the process (Hörtensteiner 2006). Both, the beginning and the rate of senescence may be affected by autonomous and environmental signals.

Environmental factors such as light (Weaver & Amasino 2001), nutrient availability, concentration of CO_2, abiotic and biotic stresses caused by disease (Sadras et al. 2000) may affect the rate of senescence. A previous work (Pic et al. 2002) showed that the sequence of certain events at macroscopic, biochemical and molecular level in pea leaf senescence were not modified in leaves of different age, or under conditions of moderate water stress. Since some of the environmental conditions that affect senescence have important effects on carbon metabolism, previous works assigned to sugar content in leaves an integrating role of environmental signals, regulating leaf senescence (Wingler et al. 2006). Reproductive growth is mentioned as a factor that usually impacts on leaf senescence, and particularly in sunflower, the lack of sinks delays the onset of senescence (Sadras et al. 2000). Control of senescence by growth of reproductive structures was not observed in *Arabidopsis thaliana* (Noodén & Penny 2001). Moreover, determining the onset of senescence is complex because there is no a "symptom" indicating this moment. Visual parameters are often used to assess these processes, but both the variation in chlorophyll content and yellowing or necrosis of leaves, are detectable long after the signalling cascade of senescence process is activated.

Senescence studies are generally based on the accumulation of messenger RNA coding for enzymes involved in degradation of structures, however, this process has a high degree of interaction between endogenous and environmental signals, involving different genes whose expression is induced or inhibited in different stages of the process (Gan & Amasino 1997). On the other hand, there are relevant studies that inversely correlate senescence with a high level of nitrogen in soil. According to these evidences a high nutritional nitrogen performance along soil profile should lead to a delay leaf senescence in sunflower, avoiding the pronounced symptoms occurred for chlorophyll content (Aguera et al. 2010).

2. Candidate gene approach to identify SAGs in sunflower

Senescence Associated Genes (SAGs) refer to genes whose expression level is up-regulated during senescence, in contrast with Senescence Down-regulated Genes (SDGs). These genes could be classified into two classes depending on their expression patterns: Class I genes are those whose expression is only activated during senescence (senescence-specific) whereas class II are those that maintain a basal level of expression during early leaf development, but this level increases when senescence begins (Gan & Amasino 1997). The expression patterns of these genes may change in response to different conditions of plant growth. Many of these genes can be shared by different regulatory pathways whereas others may belong to a particular pathway. Thus, the inactivation or overexpression of many SAGs may not exhibits significant effect, suggesting a complex regulatory network in leaf senescence process. SAGs can be grouped into several categories based on their predictive function, including macromolecular degradation and recycling, amino acid transport, metabolism, detoxification, regulatory genes, among others (Gepstein et al. 2003).

The main objective in sunflower to open new insights into the early leaf senescence process focuses in the identification and characterization of genetic sequences and metabolic pathways involved in the onset and evolution of the leaf senescence process. This aim involved the analysis of transcriptional and metabolic profiles in leaves from plants growing under different conditions that may alter the senescence rate, concomitant with studies of physiological and biochemical aspects. The specific items involved in this work include:

1. Study of the evolution of leaf area, chlorophyll and sugar content in leaf of different ages in a traditional sunflower hybrid subjected to treatments that alter the senescence under both field and greenhouse conditions.
2. Identification in public sunflower databases of gene sequences orthologous to Senescence Associated Genes (SAG) or Senescence Down-regulated Genes (SDG).
3. Identification of new candidate genes through a sunflower microarray expression analysis.
4. Verification and quantification of the expression profiles of these genes under conditions that accelerate or delay the senescence process.
5. Study of metabolic changes that occurred during the senescence process.
6. Integration of metabolic and transcriptional profile analysis and physiological variables for the detection of useful biomarkers for application in sunflower breeding.

Following a candidate genes strategy, a preliminary assay to detect putative SAGs in sunflower was achieved by selecting few candidates previously described for *Arabidopsis thaliana*, due to the fact that this was the very first model plant for which a large-scale SAG transcriptome was available (Gepstein et al. 2003). For this purpose six candidate SAGs were selected from this plant model (Moschen 2009) to search for orthologous genes in the sunflower EST database using the tblastx algorithm (Altschul et al. 1990), employing bioinformatics tools locally installed and developed. Sequences showing significant similarity parameters were selected and confirmed. Specific oligonucleotides were designed to amplify fragments of approximately 150 bp for further evaluation by quantitative PCR. In a previous study, we have reported the evaluation and identification of a panel of eight reference genes for their application to transcriptional analysis of the leaf senescence process, thus enabling the use of genuine reference genes in ongoing expression studies (Fernandez et al. 2011). Exploratory studies of senescence by qPCR comparing two treatments which affect the rate of leaf senescence were performed: water stress and head excision, relative to a control condition. Samples were taken from two leaves of different ages, leaf 15 and 25 in order to identify functional markers for this process. Two of the selected genes, a gamma vacuolar processing enzyme (AN At5g60360) (D3 gene) involved in the maturation and activation of vacuolar proteins and an aleurain protease AALP, (AN At1g18210) (D4 gene), belonging to the cystein-protease family are classified in the group of macromolecular degradation and recycling; the third gene, a calcium binding protein (AN At4g32940) (R2 gene) belongs to the group of regulatory genes (Gepstein et al. 2003). Furthermore two reference genes were evaluated against these conditions for relative expression studies, Elongation Factor 1-α (AN) and α-Tubuline, selected from a previous study of the performance of different reference genes against these experimental conditions in sunflower (Fernandez et al. 2011). Alfa tubuline (α-Tubuline) showed the most stable behavior; therefore, it was selected as internal control in further analysis of expression of these SAGs (Figure 1).

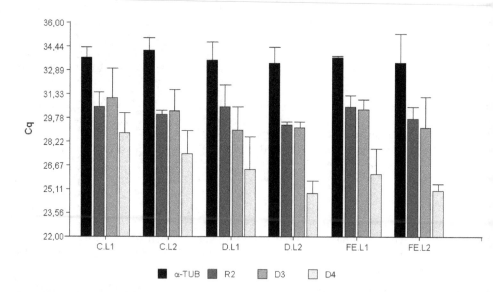

Fig. 1. Average Cq of analyzed SAGs genes normalizing against a-TUB as RG. Error bars show standard deviation (Fernandez et al. 2011).

The three selected genes did not show significant differences between the evaluated conditions at the sampling times tested (63 days post-emergence) (Table 1). It is worth noting that the target genes showed high expression levels even in controls plants with values close to the water stressed samples. Thus, these genes were probably induced by internal plant factors at an early time point, prior to the tested time in that assay. On the other hand, sampling for the incidence of head excision assessment on senescence could be consistent with an early stage of bud development in which there would be no evident differences between the two conditions (Zavaleta-Mancera et al. 1999a; Zavaleta-Mancera et al. 1999b; Thomas & Donnisson 2000).

| Treatment | Samples | SAGs genes (Gepstein et al. 2003) | | | | | | RGs genes (Fernandez et al. 2011) | | | |
| | | R2 (AN At4g32940) | | D3 (AN At5g60360) | | D4 (AN At1g18210) | | α-TUB (AN AF401481.1) | | EF-1α (AN CAA37212.1) | |
		Cq	CV	Cq	CV	Cq	CV	Cq	CV	Cq	CV
C.L1	3	30.49	2.5	31.06	5.0	28.74	3.9	33.69	1.7	30.08	2.6
C.L2	3	30.00	0.8	30.19	3.8	27.42	4.5	34.20	1.9	25.57	7.2
FE.L1	3	30.46	3.9	28.96	4.3	26.42	6.6	33.52	2.9	27.16	1.7
FE.L2	3	29.28	0.6	29.13	1.1	24.84	2.7	33.32	2.6	26.73	12.9
D.L1	3	30.45	2.1	30.31	1.7	26.10	5.1	33.67	0.2	27.80	6.7
D.L2	3	29.75	2.0	29.12	5.5	24.98	1.5	33.38	4.5	30.07	5.7

Table 1. Average Cq and CV value for R2, D3 and D4 genes and the two best ranked RGs for three biological replicates per treatment (Fernandez et al. 2011).

As a result from these analyses, the adjustment of the sampling time and frequency turns out as a highly critical point in studying gene expression profiling of candidate genes, according to the treatments on evaluation. Earlier samplings are necessary to detect the trigger moment of different candidate genes for leaf senescence process in sunflower. Considering Table 1, it is worth mentioning that relative quantification of a putative SAG would be overestimated if EF-1α (AN CAA37212.1) would have been used as a single reference gene, which reinforces the importance of normalizing against two or more experimentally validated RG when quantifying transcripts (Fernandez et al. 2011). In order to reach a wider search of new candidate genes, an additional set of new published genes were considered and their predicted functionality was evaluated with the aim to give new insights into this process. For a preliminary detection of potential SAGs, classical macromolecular degradation SAGs were discarded of our analysis because they are probably not associated with early leaf senescence, but with induced changes later in the time course of the process. In this sense, Chlorophyll-Binding Proteins (CBP) were first isolated in soybean (Guiamet et al. 1991) whereas SAGs N4 and SAG12 were detected by differential screen of Arabidopsis leaf senescence cDNA libraries (Gan & Amasino 1995; Park et al. 1998). They encode an apparent cysteine proteinase and their expression is highly senescence specific (Lohman et al. 1994; Gan & Amasino 1995; Martinez et al. 2007) mainly localized in small senescence associated vacuoles (Saeed et al. 2003; Otegui et al. 2005). However, neither SAG12 nor SEN4 match any full sequence in sunflower with a high identity score level. For this reason, a second set of candidate SAGs (OsNAC5, WRKY6, ORS1 YUCCA6, among others) (Ülker & Somssich 2004; Balazadeh et al. 2011; Kim et al. 2011; Song et al. 2011) was compared against *Helianthus annuus* unigene collection but a low score level to *Helianthus annuus* sequences was detected. Therefore, other candidate genes were added to be functionally tested for early leaf senescence in sunflower. The special case of transcription factors (TFs) as crucial regulators of gene expression by binding to distinct cis-elements, generally located in the 5' upstream regulatory regions of target genes, were specially considered to detect early senescence leaf makers (Balazadeh et al. 2008). NAC transcription factors related to senescence have been recently identified in model species and they play a relevant role in the regulation of development of leaf senescence related to programmed cell death (Olsen et al. 2005; Kim et al. 2009; Balazadeh et al. 2010; Hu et al. 2010; Nuruzzaman et al. 2010; Balazadeh et al. 2011). A single one NAC gene (AtNAP), also called NAC2 or ANAC029 (Guo & Gan 2006), has been the main one identified to control leaf senescence, although approximately 20 NAC genes in *Arabidopsis* shown high expression in senescing leaves (Guo et al. 2004; Lin & Wu 2004). ROS reagents acting as senescence stimulus were also reported within a narrow cross talk involving hormones and TFs both in natural and stress-related senescence (Rivero et al. 2007; Khanna-Chopra 2011), indicating that elevated ROS levels might be detected as a potential signal of senescence induction. Under this assumption *ORE1*, a NAC transcription factor that has been extensively studied in recent years, has been described as strongly related to leaf senescence, probably coevolving genes with ORS1 (Ooka et al. 2003). This TF can be considered a new further positive regulator of senescence in conjunction with AtNAP (Balazadeh et al. 2011), controlling leaf senescence in *Brassicaceae*. In *Arabidopsis, ORE1* mutants show a delay in leaf senescence whereas overexpression through an inductive promoter, accelerates senescence in relation to wild type plants (Balazadeh et al. 2010) and the forest tree *Populus trichocarpa* in which approximately 2,900 TFs were reported (Hu et al. 2010) and will be soon tested for sunflower candidate SAG detection. Microarray studies showed that 46% of up regulated genes in *Arabidopsis ORE1* overexpression lines, are known as senescence-associated genes, including many genes previously reported as senescence regulated, suggesting an important role in the development of the senescence process (Balazadeh et al. 2010). In wheat, it was reported that NAC TFs not only accelerate senescence but also improve nutrient remobilization by increasing protein, iron and zinc content (Uauy et al. 2006). ORE1 expression is under control of the ethylene signaling pathway and is subjected to regulation by miRNA164, being negatively regulated. When the leaf is young, miR164 transcripts remain at high levels regulating the expression of ORE1 but during the leaf aging process, its expression gradually decreases, thus increasing the expression of ORE1 (Kim et al. 2009).

In sunflower, a sequence similar ORE1 has been detected in the *Helianthus annuus* unigene collection developed at INTA (ATGC Sunflower Database: http://bioinformatica.inta.gov.ar/ATGC) with a Blast score of 96 and E-value of e-10 (Altschul et al. 1990). Expression profiles studies at different sunflower developmental stages showed a significant increase of putative ORE1 transcripts in samples close to anthesis stage, prior to the start of the first symptoms of senescence, when the critical period f grain filling has already begun (Figure 2). These results are consistent with hose observed in *Arabidopsis*, and turn this gene a potential functional marker of the progress of senescence, representing an important tool for future implications in the sunflower crop improvement (Moschen et al. 2010). In order to confirm *in-situ* the unctionality of this putative ORE1 gene in sunflower, a comparative bioinformatics analysis has been performed using the Blastx algorithm (Altschul et al. 1990), searching for proteins in the database at the National Center for Biotechnology Information NCBI (http://www.ncbi.nlm.nih.gov/), using as query the nucleotide sequence of putative sunflower ORE1. These results showed a high similarity with ORE *Arabidopsis* protein (GI 15241819) suggesting a possible role of this gene as NAC transcription factor. Moreover, searches for functional protein domains in Pfam (http://pfam.sanger.ac.uk/) revealed that main protein domain in sunflower ORE1-like gene sequence corresponds to the family of NAM transcription factors (No Apical Meristem) (pfam02365), as well as the *Arabidopsis* ORE1 sequence pfam02365. Figure 3 shows *Arabidopsis* alignments and putative sunflower ORE1 proteins against Pfam NAC domain. Others relevant *in-silico* candidates for a putative sunflower SAG are: RAV1 gene, a transcription factor whose expression is closely associated with leaf maturation and senescence (Woo et al. 2010), which has been detected with a high score level and statistically low E-value, and CAT2, a member of a small gene H_2O_2 detoxifying enzyme family, widely characterized in *Arabidopsis* (Gergoff et al. 2010; Smykowski et al. 2010), although not yet tested in sunflower.

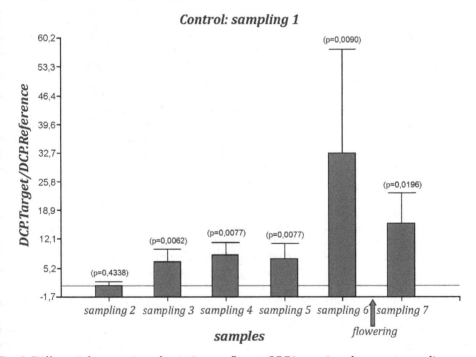

Fig. 2. Differential expression of putative sunflower *ORE1* gene in subsequent samplings, taking as control condition sampling number 1 and referred to α-TUB expression level (Moschen et al. 2010).

```
                  10        20        30        40        50        60        70        80
            ....*....|....*....|....*....|....*....|....*....|....*....|....*....|....*....|
gi 15241819    20 LPPGFRFHPTDEELITHYLKPKVF-NTFFSATAIGEVDLNKIEPWDLPW-KAKMGEKEWYFFCVDRKYPTGLRTNRATE 97
Cdd:pfam02365   1 LPPGFRFHPTDEELVVYYLKRKVLgKPLPLLEVIPEVDIYKFEPWDLPDgKAKGGDREWYFFSPRDRKYPNGSRTNRATG 80
Sunflower ORE1  1 MPPGFRFHPTDEELITHYLSNKVV-DNNFVAKAIAEVDMNKIEPWELPK-LAKLGGKEWYFFCVNDKKYPTGLRTNRATA 78


                  90       100       110       120       130
            ....*....|....*....|....*....|....*....|....*....|
gi 15241819    98 AGYWKATGKDKEIF-KGKSLVGMKKTLVFYKGRAPKGVKTNWVMHEYRLE 146
Cdd:pfam02365  81 SGYWKATGKDKPILsKGGEVVGMKKTLVFYKGRAPKGEKTDWVMHEYRLE 130
Sunflower ORE1 79 AGYWKATGKDKEIS RGKLLVGMKKXTLVFYMGRAPKGEKTNWVIHEYRLE 128
```

Fig. 3. *Arabidopsis* NAM domain and putative *sunflower* ORE1 protein alignment
(pfam02365) (http://www.ncbi.nlm.nih.gov/cdd/).

As mentioned above, the execution of the senescence process consists of multiple
interconnecting pathways which regulate and/or modulate this series of orderly steps,
therefore different transcription factors play an important role as regulators of these
pathways. Recently, a list of transcription factors that regulate leaf senescence in *Arabidopsis*
has been published (Balazadeh et al. 2008). The search for tentative orthologous genes in the
Helianthus annuus unigene collection, using Blast algorithm, led to the identification of 42
genes with a significant score value to transcription factors like NAC, MYB, WRKY, ARF
among others, some of these genes are being studied their expression patterns by qPCR.

3. Concerted gene expression studies to elucidate sunflower senescence process

Although microarray technology started a new era of high-throughput transcriptomic
analysis approximately ten years ago, starting with 8,000 printed genes by Affymetrix in
Arabidopsis thaliana (Zhu & Wang 2000) and later on scaling up to 45,000 printed genes in rice
(Jung et al. 2008) and 90,000 in *Brassica* (Trick et al. 2009), next generation sequencing (NGS)
technologies are nowadays opening a new era of even deeper understanding of genomics and
transcriptomics in different species . However, for the foreseeable future both technologies will
coexist each focusing on different tasks, or by complementing biological and value information
(Fenart et al. 2010) or by designing dedicated oligonucleotide arrays to support functional
studies on a specified pathway/developmental stage (Kusnierczyk et al. 2008; Cosio &
Dunand 2010; Ott et al. 2010). One obvious application of microarray technology is the
transcriptional profiling in species that have neither their own genome sequenced nor a
reference genome from a closely related species. For some of these species a commercial
microarray based on an existing own-design are available (Agilent, Affimetrix, Nimblegen,
etc) (Close et al. 2004; Li et al. 2008; Martinez-Godoy et al. 2008; Mascarrell-Creus et al. 2009;
Trick et al. 2009; Booman et al. 2010; Curtiss et al. 2011). Sunflower is a species that fits into this
framework, even though a genome sequence initiative is in progress (Kane et al. 2011), there is
no reference genome available. In this case, the only source of functional information is limited
to ESTs databases, which in the case of cultivated sunflower is rather extensive, more than
133,000 ESTs are publicly available (http://ncbi.nlm.nih.gov/dbEST/dbEST_summary.html)
covering libraries prepared from several lines and cultivars (Table2). However, it should also
be noted that ESTs libraries tend to be significantly contaminated with vector sequences and
chimeras, and have relatively low quality DNA information derived from the library
sequencing strategy which prioritizes obtaining a large number of single pass sequences, being
necessary to standardize a set of bioinformatics routines in order to clean and decontaminate
public raw sequences (Figure 4).

Microarrays using ESTs and full length gene sequences allowed SAGs identification during leaf senescence at the genome-wide scale in *Arabidopsis* and other plants (Lim et al. 2007). In parallel, other high-throughput system has been assayed in other species: cDNA macro and microarray were developed for sunflower to study sunflower seed development (Hewezi et al. 2006) and the response to biotic (Alignan et al. 2006), and abiotic stresses (Hewezi et al. 2006; Roche et al. 2007; Fernandez et al. 2008). This last work reported for the first time, a concerted study on gene expression in early responses to chilling and salinity using a fluorescence microarray assay based on organ-specific unigenes in sunflower. These two strategies, although useful, are limited to the analysis of a limited set of genes. Currently, the shortage of candidate genes underlying agronomically important traits represents one of the main drawbacks in sunflower molecular breeding. In this context, functional tools which allow concerted transcriptional studies, as high density oligonucleotide microarray, strongly support the discovery and characterization of novel genes. Oligonucleotide-based chips not only allow the analysis for a whole transcriptome but they are also considered more accurate than cDNA-based chips due to the reduction of manipulation steps (Larkin et al. 2005; Lai et al. 2006). The possibility to implement this technology on any custom array system like Agilent, Nimblegen, and others, has the potential to create a very useful tool for gene discovery in orphan crops (Nazar et al. 2010; Ophir et al. 2010). In addition, the use of longer probe format represents a major advantage of Agilent oligonucleotide microarrays over others technologies based on a higher stability in the presence of sequence mismatches, being consequently, more suitable for the analysis of highly polymorphic regions (Hardiman 2004).

In general, the analysis of complex biological processes based on a gene by gene approach seldom leads to limited or erroneous conclusions requiring an alternative approach based on systemic association studies. Under this assumption, new insights into molecular senescence events might be cleared up by high-resolution microarray data, for example, considering different points of leaf development (Breeze et al. 2011) or predicting putative SAGs by tissue and functional categories (Thomas et al. 2009). In our lab, a public and proprietary datasets of *H. annuus L.* ESTs have been used to create a comprehensive sunflower unigene collection. This dataset comprises 34 cDNAs libraries available from different cultivars, various tissues and anatomical parts, from plants grown at different physiological conditions.

Figure 4 describes the routines applied for the *H. annuus L* unigene collection design.

A Digital Gene Expression Profile (Audic & Claverie 1997) was assayed with the EST public data in order to detect any bias that would be pseudo-enriching the gene index by full representation of one library over another considering full public ESTs derived from public collections (Table 2). This analysis ("digi-Northern") detected that ESTs were equally represented among differential cDNA libraries, showing that the *H. annuus* unigene collection generated would be fully represented by different transcripts, lacking of a potential enrichment or overestimation among organ-specific ESTs libraries. This unigene collection was used to design the first custom sunflower oligonucleotide-based microarray based on Agilent technology as a main goal for functional genomics approaches, generated within the frame of a collaborative project involving Argentinean research sunflower groups (Sunflower PAE Consortium), Facultad de Agronomía (UBA) and the Bioinformatics facility at the Principe Felipe Institute, Valencia , España. A Chado-based database (Mungall et al.

2007) and a visualization tool call ATGC (Clavijo et al., unpublished) was developed to integrate and browse sunflower transcriptome information. Figure 5 shows the output of the ATGC interface for one functional annotated sunflower unigene.

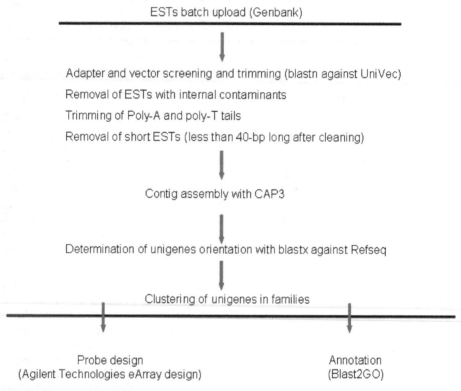

Fig. 4. Bioinformatics routines applied to design *Helianthus annuus* unigene collection (http://bionformatica.inta.gov.ar/ ATGC/).

Sunflower gene expression chip probes were designed using eArray® web application (Agilent Technologies). For this instance, two probe sets were designed: one including non-control specific probes for the sequences of sunflower unigene collection and a second control probe set consisting in 74 probes derived from 80 differentially expressed sunflower genes identified in a previously work (Fernandez et al. 2008). The latest group was used as 'Replicate Controls' with 10 replicates each. To utilize the full capacity of the microarray, probes were randomly selected to be represented in duplicate in the final design, which also included Agilent Technologies' standard panel of quality control and spike-in probes. This design was then used to manufacture microarrays using Agilent SurePrint™ Technology in the 4 x 44 format. Agilent's microarrays include the Spike-In Kit that consists of a set of 10 positive control transcripts optimized to anneal to complementary probes on the microarray, minimizing self-hybridization or cross-hybridization. This work contemplates the microarray validation through diverse differential expression analysis in order to analyze early senescence in sunflower through a classical approach and a pipeline-based methodology. Differential gene expression was also carried out using the limma package (Smyth 2004). Multiple testing adjustments of p-values was done according to Benjamini and Hochberg methodology (Benjamini & Hochberg 1995). Gene set analysis was carried out according to the Gene Ontology terms using FatiScan (Al-Shahrour et al. 2007) integrated in Babelomics suite (Al-Shahrour et al. 2005).

Library ID	Developmental stage
HaSSH	Molecular characterization of phosphorus-responsive genes in sunflower
CCF (STU)	EST sequences from several different strains/cultivars
QH-RHA 280/QH_ABCDI sunflower RHA801	shoots/hulls/flowers environmental stress/chemical induction
CHA(XYZ) common wild sunflower	girasol silvestre (wild sunflower)
HaHeaS	heart-shaped embryo vs cotyledonary embryo
HaHeaR	heart-shaped embryo
HaCotR	cotyledonary embryo
HaGlbR	globular embryo
HaDevS1	4 days after self-pollination embryo
HaDevS2	7 days after self-pollination embryo
HaDevR1	leaves
HaDevR2	terminal bud
HaDevR3	stem
HaDevR6	embryo
HaDevR5	4 days after self-pollination embryo
HaDevR8	15 days after self-pollination embryo
HaDis	unknown/cotyledons/ (Genoplante)
HaSemS4	hypocotyl
HaDpsR1	hypocotyl
HaDplR2	hypocotyl 1-5 days
HaDplR	protoplast
HaERF	embryo
HaERS	embryo
HaR	INTA: organ-specific cDNA libraries (root)
HaT	INTA: organ-specific cDNA libraries (stem)
HaEF	INTA: organ-specific cDNA libraries (early flower)
HaF	INTA: organ-specific cDNA libraries (flower)
HaH	INTA: organ-specific cDNA libraries (leaf)

Table 2. Public cDNA libraries deposited in GenBank for which *H. annuus* unigene collection was designed.

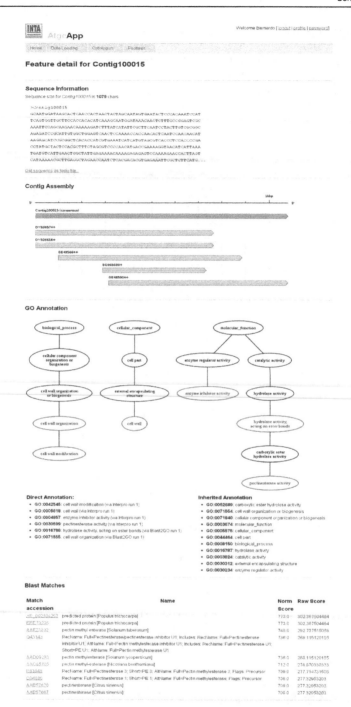

Fig. 5. ATGC view for an annotated sunflower unigene.

4. Conclusions and perspectives

Knowing the time of onset the the cascade of events that trigger senescence could determine he causes of this process and generate molecular tools to facilitate future interventions on it, iseful for application in assisted breeding of this crop with major growing oil impact in the world.

The sunflower chip, designed within a PAE Consortium made up of six laboratories and ine private company working in different areas of research and development, was validated by means of the analysis of global changes in gene expression profiles in response to water leficit as a physiological event which induces senescence, taken as a model experiment, for which reference genes have also been previously identified (Fernandez et al. 2011). This nigh-throughput transcriptome tool will allow the discovery, identification and analysis of a new set of putative SAGs for sunflower which would bring novel insights for this process. The integrated analysis of transcriptional and metabolic profiles will allow the identification of concerted regulation of distinct metabolic pathways facilitating the discovery of robust candidate genes and key metabolic pathways involved in the outbreak of the early senescence process in sunflower leaves. We expect that the integration of the information generated by this project will allow the construction of the quantitative predictive model of senescence in sunflower, under field and greenhouse conditions, which is required to nterpret the regulation of the underlying complex biological processes. There will also be practical applications in directed gene discovery for other important agronomic traits nvolving plant responses to biotic and abiotic stresses. Finally, this project will have impact based in the establishment of microarray technologies and metabolic analysis, as well as on the knowledge of appropriated statistical and bioinformatics procedures supporting functional genomics ranging from the transcriptome to the metabolome.

5. Acknowledgment

This research was supported by CONICET PIP 5788, ANPCyT/FONCYT, Préstamo BID PICT 15-32905 and PICT 0960, INTA-PE AEBIO 241001and 245001, INTA-PE AEBIO 245732, INTA-AEBI0 243532, INTA PN CER 1336 and UNMdP, AGR212, AGR260. Lic. Sebastián Moschen holds a fellowship from ANPCyT to support his PhD studies whereas Dr. PdCF, Dr. RAH, Dr. NBP are career members of the Consejo Nacional de Investigaciones Científicas y Técnicas (CONICET, Argentina) and INTA researchers.

6. References

Aguera, E., P. Cabello and P. de la Haba (2010). "Induction of leaf senescence by low nitrogen nutrition in sunflower (*Helianthus annuus*) plants." *Physiol Plant* 138(3): 256-267.

Aguirrezábal, L. A. N. A., Y. Lavaud, G. A. A. Dosio, N. Izquierdo, F. Andrade and L. González (2003). "Intercepted solar radiation during seed filling determines sunflower weight per seed and oil concentration." *Crop Science* 43: 152-161.

Al-Shahrour, F., L. Arbiza, H. Dopazo, J. Huerta-Cepas, P. Minguez, D. Montaner and J. Dopazo (2007). "From genes to functional classes in the study of biological systems." *BMC Bioinformatics* 8(1): 114.

Al-Shahrour, F., P. Minguez, J. M. Vaquerizas, L. Conde and J. Dopazo (2005). "BABELOMICS: a suite of web tools for functional annotation and analysis of groups of genes in high-throughput experiments." *Nucleic Acids Res* 33(Web Server issue): W460-464.

Alignan, M., T. Hewezi, M. Petitprez, G. Dechamp-Guillaume and L. Gentzbittel (2006). "A cDNA microarray approach to decipher sunflower (*Helianthus annuus*) responses to the necrotrophic fungus *Phoma macdonaldii*." *New Phytol* 170(3): 523-536.

Altschul, S., W. Gish, W. Miller, E. Myers and D. Lipman (1990). "Basic local alignment search tool." *Journal of Molecular Biology* 215: 403-410.

Andersen, C., J. Jensen and T. Orntoft (2004). "Normalization of real-time quantitative reverse transcription–PCR data: a model-based variance estimation approach to identify genes suited for normalization, applied to bladder and colon cancer data sets." *Cancer Res* 64: 5245–5250.

Audic, S. and J. M. Claverie (1997). "The significance of digital gene expression profiles." *Genome Research* 7(10): 986-995.

Balazadeh, S., M. Kwasniewski, C. Caldana, M. Mehrnia, M. I. Zanor, G. P. Xue and B. Mueller-Roeber (2011). "ORS1, an HO-responsive NAC transcription factor, controls senescence in Arabidopsis thaliana." *Mol Plant* 4(2): 346-360.

Balazadeh, S., D. M. Riaño-Pachón and B. Mueller-Roeber (2008). "Transcription factors regulating leaf senescence in *Arabidopsis thaliana*." *Plant Biology* 10((Suppl. 1)): 63-75.

Balazadeh, S., H. Siddiqui, A. D. Allu, L. P. Matallana-Ramirez, C. Caldana, M. Mehrnia, M.-I. Zanor, B. Köhler and B. Mueller-Roeber (2010). "A gene regulatory network controlled by the NAC transcription factor ANAC092/AtNAC2/ORE1 during salt-promoted senescence." *The Plant Journal* doi: 10.1111/j.1365-313X.2010.04151.x.

Balazadeh, S., H. Siddiqui, A. D. Allu, L. P. Matallana-Ramirez, C. Caldana, M. Mehrnia, M. I. Zanor, B. Kohler and B. Mueller-Roeber (2010). "A gene regulatory network controlled by the NAC transcription factor ANAC092/AtNAC2/ORE1 during salt-promoted senescence." *Plant J* 62(2): 250-264.

Balazadeh, S., A. Wu and B. Mueller-Roeber (2010). "Salt-triggered expression of the ANAC092-dependent senescence regulon in Arabidopsis thaliana." *Plant Signal Behav* 5(6): 733-735.

Benjamini, Y. and Y. Hochberg (1995). "Controlling the False Discovery Rate: A Practical and Powerful Approach to Multiple Testing." *Journal of the Royal Statistical Society* 57(1): 289-300.

Booman, M., T. Borza, C. Y. Feng, T. S. Hori, B. Higgins, A. Culf, D. Leger, I. C. Chute, A. Belkaid, M. Rise, A. Kurt Gamperl, S. Hubert, J. Kimball, R. J. Ouelelette, S. C. Johnson, S. Bowman and M. L. Rise (2010). "Development and Experimental Validation of a 20K Atlantic Cod (*Gadus morhua*) Oligonucleotide Microarray Based on a Collection of over 150,000 ESTs." *Mar Biotechnol* DOI: 10.1007/s10126-010-9335-6.

Breeze, E., E. Harrison, S. McHattie, L. Hughes, R. Hickman, C. Hill, S. Kiddle, Y.-s. Kim, C. A. Penfold, D. Jenkins, C. Zhang, K. Morris, C. Jenner, S. Jackson, B. Thomas, A. Tabrett, R. Legaie, J. D. Moore, D. L. Wild, S. Ott, D. Rand, J. Beynon, K. Denby, A. Mead and V. Buchanan-Wollaston (2011). "High-Resolution Temporal Profiling of Transcripts during Arabidopsis Leaf Senescence Reveals a Distinct Chronology of Processes and Regulation." *The Plant Cell Online* 23(3): 873-894.

Buchanan-Wollaston, V., S. Earl, E. Harrison, E. Mathas, S. Navabpour, T. Page and D. Pink (2003). "The molecular analysis of leaf senescence -a genomic approach." *Plant Biotechnology Journal* 1: 3-22.

Buermans, H., Y. Ariyurek, G. van Ommen, J. den Dunnen and P. 't Hoen (2010). "New methods for next generation sequencing based microRNA expression profiling." *BMC Genomics* 11(1): 716.

Cabello, P., E. Agüera and P. De la Haba (2006). "Metabolic changes during natural ageing in sunflower (*Helianthus annuus*) leaves: expression and activity of glutamine synthetase isoforms are regulated differently during senescence." *Physiol Plant* 128: 175-185.

Close, T. J., S. I. Wanamaker, R. A. Caldo, S. M. Turner, D. A. Ashlock, J. A. Dickerson, R. A. Wing, G. J. Muehlbauer, A. Kleinhofs and R. P. Wise (2004). "A New Resource for Cereal Genomics: 22K Barley GeneChip Comes of Age." *Plant Physiology* 134: 960-968.

Conesa, A., S. Gotz, J. M. Garcia-Gomez, J. Terol, M. Talon and M. Robles (2005). "Blast2GO: a universal tool for annotation, visualization and analysis in functional genomics research." *Bioinformatics* 21(18): 3674-3676.

Cosio, C. and C. Dunand (2010). "Transcriptome analysis of various flower and silique development stages indicates a set of class III peroxidase genes potentially involved in pod shattering in Arabidopsis thaliana." *BMC Genomics* 11(1): 528.

Curtiss, J., L. Rodriguez-Uribe, J. M. Stewart and J. Zhang (2011). "Identification of differentially expressed genes associated with semigamy in Pima cotton (Gossypium barbadense L.) through comparative microarray analysis." *BMC Plant Biol* 11: 49.

De la Vega, A., M. A. Cantore, N. N. Sposaro, N. Trapani, M. Lopez Pereira and A. J. Hall (2011). "Canopty stay green and yield in non stressed sunflower." *Fields Crop Researchs* 121: 175-185.

Dezar, C. A., G. M. Gago, D. H. Gonzalez and R. L. Chan (2005). "Hahb-4, a sunflower homeobox-leucine zipper gene, is a developmental regulator and confers drought tolerance to *Arabidopsis thaliana* plants." *Transgenic Res* 14(4): 429-440.

Dhahbi, J. M., H. Atamna, D. Boffelli, W. Magis, S. R. Spindler and D. I. K. Martin (2011). "Deep Sequencing Reveals Novel MicroRNAs and Regulation of MicroRNA Expression during Cell Senescence." *PLoS ONE* 6(5): e20509.

Dosio, G. A., H. Rey, J. Lecoeur, N. G. Izquierdo, L. A. Aguirrezabal, F. Tardieu and O. Turc (2003). "A whole-plant analysis of the dynamics of expansion of individual leaves of two sunflower hybrids." *J Exp Bot* 54(392): 2541-2552.

Dosio, G. A. A. and L. A. N. Aguirrezábal (2004). Variaciones del rendimiento en girasol. Identificando las causas. *Revista Agromercado*, Cuadernillo de girasol. 90: 7-10.

Dosio, G. A. A., L. A. N. A. Aguirrezábal, F. H. Andrade and P. V. R. (2000). "Solar radiation intercepted during seed filling and oil production in two sunflower hybrids." *Crop Science* 40(1637-1644).

Espinoza, C., C. Medina, S. Somerville and P. Arce-Johnson (2007). "Senescence-associated genes induced during compatible viral interactions with grapevine and Arabidopsis." *Journal of Experimental Botany* 58(12): 3197-3212.

Ewing, R. M. and J. M. Claverie (2000). "EST databases as multi-conditional gene expression datasets." *Pac Symp Biocomput*: 430-442.

Fenart, S., Y.-P. Assoumou Ndong, J. Duarte, N. Rivière, J. Wilmer, O. van Wuytswinkel, A. Lucau, E. Cariou, G. Neutelings, L. Gutierrez, B. Chabbert, X. Guillot, R. Tavernier, S. Hawkins and B. Thomasset (2010). "Development and validation of a flax (*Linum usitatissimum L.*) gene expression oligo microarray." *BMC Genomics* 11(592).

Fernandez, P., J. Di Rienzo, L. Fernandez, H. Hopp, N. Paniego and H. R.A. (2008). "Transcriptomic identification of candidate genes involved in sunflower responses to chilling and salt stresses based on cDNA microarray analysis." *BMC Plant Biology* 8(11).

Fernandez, P., J. Di Rienzo, S. Moschen, D. GAA, L. Aguirrezabal, H. Hopp, N. Paniego and H. R.A. (2011). "Comparison of predictive methods and biological validation for qPCR reference genes in sunflower leaf senescence transcript analysis." *Plant Cell Report* 30(1): 63-74.

Fernandez, P., N. Paniego, S. Lew, H. E. Hopp and R. A. Heinz (2003). "Differential representation of sunflower ESTs in enriched organ-specific cDNA libraries in a small scale sequencing project." *BMC Genomics* 4(1): 40.

Gan, S. and R. M. Amasino (1995). "Inhibition of leaf senescence by autoregulated production of cytokinin." *Science* 270(5244): 1986-1988.

Gan, S. and R. M. Amasino (1997). "Making sense of senescence." *Plant Physiology* 113: 313-319.

Gepstein, S., G. Sabehi, M.-J. Carp, T. Hajouj, M. F. O. Nesher, I. Yariv, C. Dor and M. Bassani (2003). "Large-scale identification of leaf senescence-associated genes." *The Plant Journal* 36: 629-642.

Gergoff, G., A. Chaves and C. G. Bartoli (2010). "Ethylene regulates ascorbic acid content during dark-induced leaf senescence." *Plant Science* 178: 207-212.

Guiamet, J. J., E. Schwartz, E. Pichersky and L. D. Nooden (1991). "Characterization of Cytoplasmic and Nuclear Mutations Affecting Chlorophyll and Chlorophyll-Binding Proteins during Senescence in Soybean." *Plant Physiol* 96(1): 227-231.

Guo, Y., Z. Cai and S. Gan (2004). "Transcriptome of Arabidopsis leaf senescence." *Plant, Cell and Environment* 27: 521-549.

Guo, Y. and S. Gan (2006). "AtNAP, a NAC family transcription factor, has an important role in leaf senescence." *The Plant Journal* 46(4): 601-612.

Hardiman, G. (2004). "Microarray platforms--comparisons and contrasts." *Pharmacogenomics* 5(5): 487-502.

Hewezi, T., M. Leger, W. El Kayal and L. Gentzbittel (2006). "Transcriptional profiling of sunflower plants growing under low temperatures reveals an extensive down-regulation of gene expression associated with chilling sensitivity." *J Exp Bot* 57(12): 3109-3122.

Hewezi, T., M. Petitprez and L. Gentzbittel (2006). "Primary metabolic pathways and signal transduction in sunflower (*Helianthus annuus L.*): comparison of transcriptional profiling in leaves and immature embryos using cDNA microarrays." *Planta* 223(5): 948-964.

Hörtensteiner, S. (2006). "Chlorophyll degradation during senescence." *The Annual Review of Plant Biology* 57: 55-77.

Hu, R., G. Qi, Y. Kong, D. Kong, Q. Gao and G. Zhou (2010). "Comprehensive Analysis of NAC Domain Transcription Factor Gene Family in *Populus trichocarpa.*" *BMC Plant Biology* 10(145).

Jobit, C., A. Boisson, E. Gout, C. Rascle, M. Feyre, P. Cotton and R. Bligny (2007). "Metabolic processes and carbon nutrient exchanges between host and pathogen sustain the disease development during sunflower infection by *Sclerotinia sclerotiorum.*" *Planta* 226: 251-265.

Jung, K. H., C. Dardick, L. E. Bartley, P. Cao, J. Phetsom, P. Canlas, Y. S. Seo, M. Shultz, S. Ouyang, Q. Yuan, B. C. Frank, E. Ly, L. Zheng, Y. Jia, A. P. Hsia, K. An, H. H. Chou, D. Rocke, G. C. Lee, P. S. Schnable, G. An, C. R. Buell and P. C. Ronald (2008). "Refinement of light-responsive transcript lists using rice oligonucleotide arrays: evaluation of gene-redundancy." *PLoS ONE* 3(10): e3337.

Kane, N. C., N. Gill, M. J. King, J. E. Bowers, H. Berges, J. Gouzy, E. Bachlava, N. B. Langlade, Z. Lai, M. Stewart, J. M. Burke, P. Vincourt, S. J. Knapp and L. H. Rieserberg (2011). "Progress towards a reference genome for sunflower." *Botany* 89: 429-437.

Khanna-Chopra, R. (2011). "Leaf senescence and abiotic stresses share reactive oxygen species-mediated chloroplast degradation." *Protoplasma.*

Kim, J. H., P. O. Lim and H. G. Nam (2007). Molecular regulation of leaf senescence. "*Senescence Process in Plants*". S. Gan. Ithaca, Blackwell Publishing.

Kim, J. H., H. R. Woo, J. Kim, P. Ok Lim, I. C. Lee, S. H. Choi, D. E. Hwang and H. Gil Nam (2009). "Trifurcate Feed-Forward Regulation of Age-Dependent Cell Death Involving miR164 in Arabidopsis." *Science* 323(5917): 1053-1057.

Kim, J. I., A. S. Murphy, D. Baek, S. W. Lee, D. J. Yun, R. A. Bressan and M. L. Narasimhan (2011). "YUCCA6 over-expression demonstrates auxin function in delaying leaf senescence in Arabidopsis thaliana." *J Exp Bot* 62(11): 3981-3992.

Kusnierczyk, A., P. Winge, T. S. Jorstad, J. Troczynska, J. T. Rossiter and A. M. Bones (2008). "Towards global understanding of plant defence against aphids--timing and dynamics of early Arabidopsis defence responses to cabbage aphid (Brevicoryne brassicae) attack." *Plant Cell Environ* 31(8): 1097-1115.

Lai, Z., B. L. Gross, Y. Zou, J. Andrews and L. H. Rieseberg (2006). "Microarray analysis reveals differential gene expression in hybrid sunflower species." *Mol Ecol* 15(5): 1213-1227.

Larkin, J. E., B. C. Franc, H. Gavras, R. Sultana and J. Quackenbush (2005). "Independence and reproducibility across microarray platforms." *Nat. Methods* 2(5): 337-344.

Li, X., H.-I. Chiang, J. Zhu, S. E. Dowd and H. Zhou (2008). "Characterization of a newly developed chicken 44K Agilent microarray." *BMC Genomics* 9:60.

Lim, P., H. Kim and H. Nam (2007). "Leaf Senescence." *Annual Rev. Plant Biol.* 58: 115-136.

Lim, P. O., H. R. Woo and H. G. Nam (2003). "Molecular genetics of leaf senescence in *Arabidopsis.*" *Trends in Plant Science* 8: 272-278.

Lin, J.-F. and S.-H. Wu (2004). "Molecular events in senescing Arabidopsis leaves." *The Plant Journal* 39(4): 612-628.

Lohman, K. N., S. Gan, M. C. John and R. M. Amasino (1994). "Molecular analysis of natural leaf senescence in Arabidopsis thaliana." *Physiol Plant* 92(2): 322-328.

Manavella, P. A., A. L. Arce, C. A. Dezar, F. Bitton, J. P. Renou, M. Crespi and R. L. Chan (2006). "Cross-talk between ethylene and drought signalling pathways is mediated by the sunflower Hahb-4 transcription factor." *Plant J* 48(1): 125-137.

Manavella, P. A., C. A. Dezar, G. Bonaventure, I. T. Baldwin and R. L. Chan (2008). "HAHB4, a sunflower HD-Zip protein, integrates signals from the jasmonic acid and ethylene pathways during wounding and biotic stress responses." *The Plant Journal* 56(3): 376-388.

Martinez-Godoy, M. A., N. Mauri, J. Juarez, M. Carmen Marques, J. Santiago, J. Forment and J. Gadea (2008). "A genome-wide 20K citrus microarray for gene expression analysis." *BMC Genomics* 9(318).

Martinez, D. E., C. G. Bartoli, V. Grbic and J. J. Guiamet (2007). "Vacuolar cysteine proteases of wheat (*Triticum aestivum L.*) are common to leaf senescence induced by different factors." *J Exp Bot* 58(5): 1099-1107.

Mascarrell-Creus, A., J. Cañizares, J. Vilarrasa-Blasi, S. Mora-Garcia, J. Blanca, D. Gonzalez-Ibeas, M. Saladié, C. Roig, W. Deleu, B. Picó-Silvent, N. López-Bigas, M. A. Aranda, J. Garcia-Mas, F. Nuez, P. Puigdomènech and A. Caño-Delgado (2009). "An oligo-based microarray offers novel transcriptomic approaches for the analysis of pathogen resistance and fruit quality traits in melon (*Cucumis melo L.*)." *BMC Genomics* 10(467).

Moschen, S. (2009). "Identificación y caracterización de genes asociados a la senescencia tempana en girasol". Tesis de Grado. Lic. en Genética.

Moschen, S., P. Fernandez, N. Paniego and R. A. Heinz (2010). ""Análisis de los perfiles de expresión de factores de transcripción NAC asociados a la senescencia foliar en girasol (*Helianthus annuus L.*)"." *Reunión Argentina de Fisiología Vegetal*.

Mungall, C. J., D. B. Emmert and T. F. Consortium (2007). "A Chado case study: an ontology-based modular schema for representing genome-associated biological information." *Bioinformatics* 23(13): i337-i346.

"National Center for Biotechnology Information (NCBI)." *http://www.ncbi.nlm.nih.gov/*.

Nazar, N. R., P. Chen, D. Dean and J. Robb (2010). "DNA Chip Analysis in Diverse Organisms with Unsequenced Genomes." *Mol. Biotechnol* 44: 8-13.

Noodén, L. and J. Penny (2001). "Correlative controls of senescence and plant death in *Arabidopsis thaliana* (Brassicaceae)." *J. Exp. Bot.* 52: 2151-2159.

Noodén, L. D. (1988). The phenomena of senescence and aging. . *Senescence and Aging in Plants*. A. C. L. L.D. Noodén. San Diego, Academic: 1-50.

Noodén, L. D., J. J. Guiamet and J. Isaac (1997). "Senescence mechanisms." *Physiologia plantarum* 101: 746-753.

Noodén, L. D. and A. C. Leopold (1987). "Phytohormones and the endogenous regulation of senescence and abscission". "*Phytohormones and related compounds: A comprehensive treatise*". Amsterdam, Elsevier: 329-370.

Nooden, L. D. and Z. C. Leopold (1988). "The phenomena of senescence and aging. In "Senescence and aging in plants"." *Academic Press*. San Diego CA, USA: 1-50.

Nuruzzaman, M., R. Manimekalai, A. M. Sharoni, K. Satoh, H. Kondoh, H. Ooka and S. Kikuchi (2010). "Genome-wide analysis of NAC transcription factor family in rice." *Gene* 465(1-2): 30-44.

Olsen, A. N., H. A. Ernst, L. L. Leggio and K. Skriver (2005). "NAC transcription factors: structurally distinct, functionally diverse." *Trends Plant Sci* 10(2): 79-87.

Ooka, H., K. Satoh, K. Doi, T. Nagata, Y. Otomo, K. Murakami, K. Matsubara, N. Osato, J. Kawai, P. Carninci, Y. Hayashizaki, K. Suzuki, K. Kojima, Y. Takahara, K. Yamamoto and S. Kikuchi (2003). "Comprehensive analysis of NAC family genes in Oryza sativa and Arabidopsis thaliana." *DNA Res* 10(6): 239-247.

Ophir, R., R. Eshed, R. Harel-Beja, G. Tzuri, V. Portnoy, Y. Burger, S. Uliel, N. Katzir and A. Sherman (2010). "High-throughput marker discovery in melon using a self-designed oligo microarray." *BMC Genomics* 11: 269.

Otegui, M. S., Y. S. Noh, D. E. Martinez, M. G. Vila Petroff, L. A. Staehelin, R. M. Amasino and J. J. Guiamet (2005). "Senescence-associated vacuoles with intense proteolytic activity develop in leaves of Arabidopsis and soybean." *Plant J* 41(6): 831-844.

Ott, H., C. Schroder, M. Raulf-Heimsoth, V. Mahler, C. Ocklenburg, H. F. Merk and J. M. Baron (2010). "Microarrays of recombinant Hevea brasiliensis proteins: a novel tool for the component-resolved diagnosis of natural rubber latex allergy." *J Investig Allergol Clin Immunol* 20(2): 129-138.

Paniego, N., R. Heinz, P. Fernandez, P. Talia, V. Nishinakamasu and H. Hopp (2007). Sunflower. *Genome Mapping and Molecular Breeding in Plants*. C. Kole. Berlin Heidelberg, Springer-Verlag. 2: 153-177.

Park, J. H., S. A. Oh, Y. H. Kim, H. R. Woo and H. G. Nam (1998). "Differential expression of senescence-associated mRNAs during leaf senescence induced by different senescence-inducing factors in Arabidopsis." *Plant Mol Biol* 37(3): 445-454.

Peluffo, L., V. Lia, C. Troglia, C. Maringolo, N. Paniego, R. A. Heinz and F. Carrari (2010). "Metabolic profiles of sunflower genotypes with contrasting response to *Sclerotinia sclerotiorum* infection." *Phytochemistry*.

Pic, E., B. Teyssandier de la Serve, F. Tardieu and O. Turc (2002). "Leaf senescence induced by mild water deficit follows the same sequence of macroscopic, biochemical, and molecular events as monocarpic senescence in pea." *Plant Physiology* 128: 236-246.

Rivero, R. M., M. Kojima, A. Gepstein, H. Sakakibara, R. Mittler, S. Gepstein and E. Blumwald (2007). "Delayed leaf senescence induces extreme drought tolerance in a flowering plant." *Proc Natl Acad Sci U S A* 104(49): 19631-19636.

Roche, J., T. Hewezi, A. Bouniols and L. Gentzbittel (2007). "Transcriptional profiles of primary metabolism and signal transduction-related genes in response to water stress in field-grown sunflower genotypes using a thematic cDNA microarray." *Planta* DOI 10.1007/s00425-007-0508-0.

Sadras, V. O., L. Echarte and A. F. H. (2000). "Profiles of Leaf Senescence During Reproductive Growth of Sunflower and Maize." *Annals of Botany* 85(2): 187-195.

Sadras, V. O., M. Ferreiro, F. Gutheim and A. G. Kantolic (2000). Desarrollo fenológico y su respuesta a temperatura y fotoperíodo. *Bases para el manejo del maíz, el girasol y la soja*. I. B. F. d. C. A. UNMP: 29-60.

Sadras, V. O., F. Quiroz, L. Echarte, E. A. and P. V. R. (2000). "Effect of *Verticillium dahliae* on Photosynthesis, Leaf Expansion and Senescence of Field-grown Sunflower." *Annals of Botany* 86: 1007-1015.

Saeed, A., V. Sharov, J. White, J. Li, W. Liang, N. Bhagabati, J. Braisted, M. Klapa, T. Currier, M. Thiagarajan, A. Sturn, M. Snuffin, A. Rezantsev, D. Popov, A. Ryltsov, E. Kostukovich, I. Borisovsky, Z. Liu, A. Vinsavich, V. Trush and J. Quackenbush (2003). "TM4: a free, open-source system for microarray data management and analysis." *Biotechniques* 34(2): 374-378.

Shahri, W. (2011). "Senescence: Concepts and Symptoms." *Asian Journal of Plant Sciences* 10(1): 24-28.

Smykowski, A., P. Zimmermann and U. Zentgraf (2010). "G-Box Binding Factor1 Reduces CATALASE2 Expression and Regulates the Onset of Leaf Senescence in Arabidopsis." *Plant Physiol* 153(3): 1321-1331.

Smyth, G. (2004). "Linear models and empirical bayes methods for assessing differential expression in microarray experiments." *Stat Appl Genet Mol Biol* 3: Article3.

Song, S.-Y., Y. Chen, J. Chen, X.-Y. Dai and W.-H. Zhang (2011). "Physiological mechanisms underlying OsNAC5-dependent tolerance of rice plants to abiotic stress." *Planta* 234(2): 331-345.

Thomas, H. and I. Donnisson (2000). "Back from the brink: plant senescence and its reversibility." *Symp Soc Exp Biol.* 52: 149-162.

Thomas, H., L. Huang, M. Young and H. Ougham (2009). "Evolution of plant senescence." *BMC Evol Biol* 9: 163.

Trick, M., F. Cheung, N. Drou, F. Fraser, E. K. Lobenhofer, P. Hurban, A. Magusin, C. D. Town and I. Bancroft (2009). "A newly-developed community microarray resource for transcriptome profiling in *Brassica* species enables the confirmation of *Brassica*-specific expressed sequences." *BMC Plant Biology* 9(50).

Uauy, C., A. Distelfeld, T. Fahima, A. Blechl and J. Dubcovsky (2006). "A NAC Gene regulating senescence improves grain protein, zinc, and iron content in wheat." *Science* 314(5803): 1298-1301.

Ülker, B. and E. I. Somssich (2004). "WRKY transcription factors: from DNA binding towards biological function." *Current Opinion in Plant Biology* 7: 491-498.

Weaver, M. and R. Amasino (2001). "Senescence is induced in individually darkened *Arabidopsis* leaves, but inhibited in whole darkened plants." *Plant Physiology* 127: 876-886.

Wingler, A., S. Purdy, J. A. MacLean and N. Pourtau (2006). "The role of sugars in integrating environmental signals during the regulation of leaf senescence." *Journal of Experimental Botany* 57(2): 391-399.

Woo, H. R., J. H. Kim, J. Kim, U. Lee, I. J. Song, H. Y. Lee, H. G. Nam and P. O. Lim (2010). "The RAV1 transcription factor positively regulates leaf senescence in Arabidopsis." *J Exp Bot* 61(14): 3947-3957.

Zavaleta-Mancera, H. A., B. J. Thomas, H. Thomas and I. M. Scott (1999a). "Regreening of senescent Nicotiana leaves: II Redifferentiation of plastids." *Journal of Experimental Botany* 50(340): 1683-1689.

Zavaleta-Mancera, H. A., B. J. Thomas, H. Thomas and I. M. Scott (1999b). "Regreening of senescent Nicotiana leaves: I Reappearance of NADPH-protochlorophyllide oxidoreductase and light-harvesting chlorophyll a/b-binding protein." *Journal of Experimental Botany* 50(340): 1677-1682.

Zhu, T. and X. Wang (2000). "Large-scale profiling of the Arabidopsis transcriptome." *Plant Physiol* 124(4): 1472-1476.

7

Plant Ageing,
a Counteracting Agent to Xenobiotic Stress

David Delmail[1,2] and Pascal Labrousse[2]
[1]University of Rennes 1, Lab. of Pharmacognosy & Mycology,
UMR CNRS 6226 SCR/PNSCM, Rennes
[2]University of Limoges, Lab. of Botany & Cryptogamy,
GRESE EA 4330, Limoges
France

1. Introduction

A xenobiotic can be defined as any chemical or other substance that is not normally found in the ecosystems or that is present at concentrations harmful to all biological organisms. This general definition could be applied to anthropogenic and naturally occurring constituents. Organic contaminants can include pesticides, solvents and petroleum products. Inorganic xenobiotics include heavy metals, nonmetals, metalloids, radionuclides and simple soluble salts (Schwab, 2005).

Indeed, after absorption in plant cell, these toxics induce a broad range of disturbances like competition between elements. But the main effect remains the oxidative stress which disrupts many physiological pathways. Reactive oxygen species which initiate the oxidation are produced through several mechanisms in all cell compartments (Delmail et al., 2009; Thompson et al., 1987). Some reactive oxygen species are less deleterious to the plant cell than others but they can act as initiator of the production of more toxic compounds (Delmail et al., 2011c, 2011d).

To prevent from the production of reactive oxygen species, plants can use the senescence process to eliminate the xenobiotics from their organisms. Toxic compounds like radionuclides can be sequestrated by metallothioneins preferentially in vacuoles of specific organs like trichomes and old leaves. Indeed, morphological structures as non-glandular trichomes which are not implied in any physiological process, could store many xenobiotics. Moreover, potentially abscised organs as mature leaves are used to eliminate toxics from the living parts to protect the young organs from any disturbance of the photosynthetic pathways (Delmail et al., 2011c, 2011d).

This chapter will focus on xenobiotics having anthropogenic origins and will address organic and inorganic xenobiotics. Moreover, the origin of the oxidative stress induced by the xenobiotic assimilation, its consequence on the ageing of morphological, physiological and cellular patterns, as well as the functioning of antioxidant pathways, the implication of scavengers and the role of the senescence in reducing the oxidative disturbance, will be discussed in this chapter.

2. The xenobiotics

A xenobiotic (from the Greek *xenos* "stranger" and *biotic* "related to living beings") is a biological (Qiu et al., 2002), physical (Sacco et al., 2004) or chemical disturbance which above a certain degree, and in certain environmental conditions, could lead to toxic effects on a part or the whole ecosystem. It implies that a xenobiotic acts as a pollutant or a contaminant of one or several compartments of the natural environments (atmosphere, lithosphere and hydrosphere) and of biological organisms among the biosphere. This compound disrupts the ecosystem functioning above the limit of tolerance.

The pollution introduced directly or indirectly by humans in all natural compartments, could have prejudicial consequences on its own species and others, on biological resources, on climates and on infrastructures (Delmail, 2007). This impact depends on the type of pollution as it could be distinguished the pollution of proximity and the regional/global pollution (Delmail et al., 2011a; Ritter et al., 2002). The first one is constituted by factory smokes, fumes, sewer gas, etc. and it is directly produced by an anthropogenic source. The second one results from more complex and diverse physicochemical phenomenon (e.g. ozone synthesis in troposphere, acid rains, greenhouse effect).

The xenobiotics could be classified according to their nature (solid, liquid, gas, mineral, organic), their radiation (X, ultraviolet, infrared, radioactivity) and their origin (natural, synecological, autoecological, chemical, industrial) (Fig. 1). They may be also distinguished depending on their environmental targets (air, soil, and water), their biological targets (e.g. plants, fungi, mammals, invertebrates) and their cytotoxicity (e.g. cell types, organites). Their mode of action brings also information as some xenobiotics have an acute (death) or chronic toxicity (e.g. carcinogenesis, mutagenesis), or synergistic effect on organisms. They could be toxic at infinitesimal concentrations (micropollutants) or at a more concentrated range (macropollutants) (Delmail et al., 2010; Delmail et al., 2011b). Moreover, their effects have different duration of action: they could be degradable or persistent, or have a half-life like radioelements from several microseconds to many thousands of years.

Ecological exposure to environmental stressors occurs when a xenobiotic in a form that is bioavailable, reaches an organism. In order to be bioavailable, a xenobiotic must reach a location on or in an organism where it can cause an effect. The notion of phytoavailability defines the fraction of a bioavailable compound which could be absorbed by roots (Hinsinger et al., 2005).

The phytoavailability of xenobiotics is strongly correlated to the concentrations of contaminant species that occurred in the natural environments (Kabata-Pendias & Pendias, 2000). It is also linked to the physicochemical properties of the environment, the plant taxon and the xenobiotic considered. Thus, the phytoavailability is dependent from several parameters allowing the transfer from aerial, solid or aqueous phase to the plant: the availability (or chemical mobility), the accessibility (or physical mobility) and the assimilation (or biological mobility) (Hinsinger et al., 2005).

The xenobiotics could be observed under free forms depending on environmental conditions, but in many cases they may interact with different elements from the environments which will have an influence on its behavior. They could be included in primary minerals originated from the rock crystallization, or in secondary minerals developed from primary ones after oxidation and hydrolysis processes. Moreover, they may be adsorbed on the organic matter in soils and waters, or on microorganisms (which could also absorb and accumulate them). Associations between colloids and xenobiotics could be observed in all natural compartments as a colloidal system may be solid, liquid or gaseous.

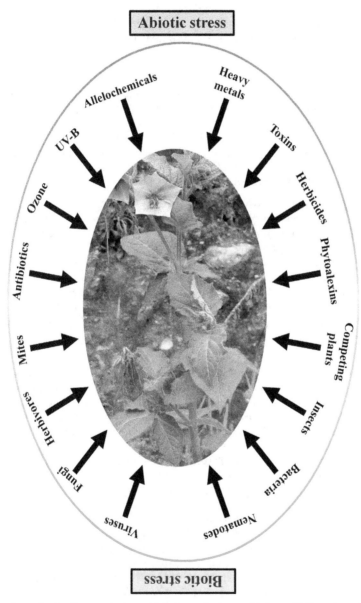

Fig. 1. Overview of selective environmental stress factors which could threaten plants. Biotic and abiotic environmental selective factors are considered.

3. The reactive oxygen species

All plants use the dioxygen as a source of energy for their development. However, this aerobic process could lead to the production of reactive oxygen species which are diversified chemically reactive molecules containing oxygen. Reactive oxygen species are a natural byproduct of the metabolism and play important roles in homeostasis and cell signaling. However, under environmental stress, their levels can increase dramatically which lead to disruptions and damages in cell compartments.

3.1 Diversity and toxicity

An uncompleted reduction of the dioxygen through cytochroms from the respiratory chain implies the production of reactive oxygen species as singlet oxygen ($^1O_2$) and superoxide radical ($O_2{}^{\bullet-}$) which leads to the synthesis of hydroxyl radical ($^{\bullet}OH$), hydroperoxyl radical ($^{\bullet}O_2H$) and hydrogen peroxide (H_2O_2) (Fig. 2). The radicals alkoxyl ($RO^{\bullet}$) and peroxyl ($RO_2{}^{\bullet}$) are the consequence of the peroxidation of membrane phospholipids (or lipoperoxidation) by the previous reactive oxygen species (Fig. 3) (Edreva, 2005; Lagadic et al., 1997; Li et al., 1994; Thompson et al., 1987).

At the same time, the photosynthetic electron transport chains could product high concentrations of reactive oxygen species. Indeed, the electrons tetravalently reduce the intracellular oxygen to water. But, some electrons could leak from many sites along the electron transport chain, resulting in a univalent reduction of dioxygen to form the extremely reactive superoxide radical which can dismutate to form hydrogen peroxide (Alscher et al., 2002). This last reaction is spontaneous or catalyzed by one of the superoxide dismutases (Fig. 2) depending on the cell compartment where the reaction occurs: manganese-superoxide dismutase (mitochondria, peroxisome), iron-superoxide dismutase (chloroplast) or copper/zinc-superoxide dismutase (chloroplast, cytosol) (Fornazier et al., 2002; Gill & Tuteja, 2010; Pereira et al., 2002).

The hydrogen peroxide is not a free radical due to all its matched electrons. However, it has a strong toxicity potential: it has a long lifespan and a high diffusibility far from its synthesis site. Indeed, it could pass through biological membranes via aquaporins as it presents a chemical structure close to water (Bienert et al., 2006, 2007; Parent et al., 2008). The concentration of this oxidative compound is regulated by antioxidant enzymes like the ascorbate peroxidase, the catalase or the glutathione peroxidase (Fig. 2). These proteins use the nicotinamide adenine dinucleotide phosphate (NADPH) produced during the photosynthesis in chloroplasts for their functioning (Fig. 2). However, the reactive oxygen species could disrupt the photosynthetic electron transport chains in thylakoid membranes and some electrons are deflected. Without a normal synthesis of NADPH, plants use a cytosolic secondary catabolism pathway to produce it, the pentose phosphate pathway (Fig. 4) (Delmail, 2011; Kruger & von Schaewen, 2003). The hydrogen peroxide could be also produced through the bivalent reduction of the oxygen in presence of oxidases like the peroxisomal glycolate oxidase or the amine oxidase (Parent et al., 2008). The toxicity of hydrogen peroxide is also linked to its implication in the synthesis of the hydroxyl and hydroperoxyl radicals through the Haber-Weiss and Fenton reactions (Fig. 2). Like their reactive-oxygen-species mother, these short-lifespan radicals are very diffusive through biological membranes and they could disturb and affect all organites and cell compartments. They are also mainly implied in the lipoperoxidation (Fig. 3) (Edreva, 2005 ; Lagadic et al.,

1997). The produced fatty-acid radical then reacts with molecular oxygen, thereby creating a peroxyl fatty acid radical. This last one reacts with another phospholipid, producing a new radical and a lipid peroxide, or a cyclic peroxide if it reacts with itself. This cycle continues as a chain reaction mechanism (Schaich, 2005). This process ends up when two radicals react and produce a non-radical compound. It happens when the concentrations of radicals is high enough. Living organisms have evolved different molecules that speed up termination by catching the reactive oxygen species (Paramesha et al., 2011). Among such antioxidants, the most important are the scavengers mainly constituted with α-tocopherol (or vitamin E) and carotenoids (β-caroten, xantophylls) (Figs. 2 and 3) (Delmail et al., 2011c, 2011d).

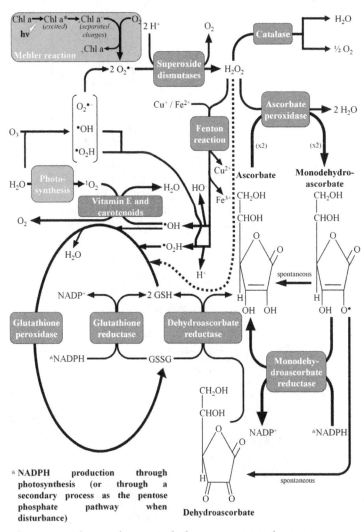

Fig. 2. Main plant antioxidant pathways including enzymes and scavengers based on Delmail (2011)).

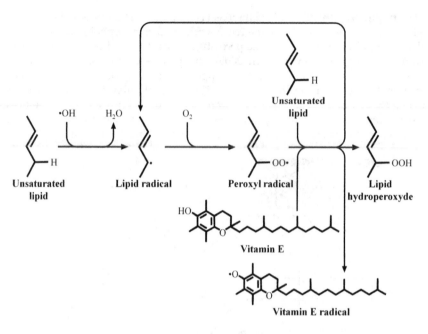

Fig. 3. Mechanisms of lipid peroxidation in biological membranes. The produced peroxyl radicals could react either with another lipid to supply the lipoperoxidative chain reaction mechanism or with a scavenger like the vitamin E which disrupts and stops the oxidative process.

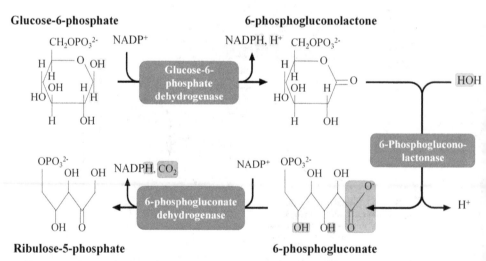

Fig. 4. Reactions of NADPH synthesis through the oxidative phase of the pentose phosphate pathway of plants (based on Delmail (2011)).

Considering all these elements, the reactive oxygen species are considered as phytotoxic compounds. However, it is currently admitted that their synthesis, in relation to the respiratory and photosynthetic metabolisms, plays an essential role in the life and the death of plant cells. Indeed, they could play an alternative role and act as cell signalization molecules to establish some defense mechanisms towards a xenobiotic stress (Parent et al., 2008).

3.2 Role in cell death and protection of living parts

The reactive oxygen species are known for their importance in the plant responses towards environmental disturbances. Several symptoms like necrosis (Fig. 5), are the consequences of a high oxidative-compound accumulation and a disturbance of cell homeostasis.

Fig. 5. Leaf of *Tradescantia sp.* (A) and *Begonia sp.* (B) with symptoms of photosynthetic-pigment oxidation and cell necrosis.

This phenomenon is due to an oxidation of photosynthetic pigments in chlorophyllian organs and to the death of isolated cells or groups of cells in many plant tissues. Despite that reactive oxygen species could be produced in normal conditions, the increase of their concentrations in plants is often linked to xenobiotics (Parent et al., 2008). For example, an increase of hydrogen peroxide is observed in peroxisomes of the aquatic macrophyte *Myriophyllum alterniflorum* after an exposition to cadmium chloride from 0.5 to 10 μg.l^{-1} (Delmail, 2011). Moreover, the oxidative stress generated by this reactive oxygen species is all the more important during the 2-3 weeks of contamination that the heavy-metal concentration is high. Indeed, the activity of the catalase is higher during a longer period when the toxicity increases (Delmail, 2011).

It could be also noted that many of the symptoms due to the xenobiotics stress are amplified by the presence of reactive oxygen species. In the same species, when *M. alterniflorum* is contaminated with copper sulphate from 5 to 100 μg.l^{-1}, an increase of the catalase activity is observed up to 25 μg.l^{-1} to reduce this reactive oxygen species into water (Delmail, 2011). Beyond this toxicity limit, the intensity of the enzymatic activity decreases due to a disruption of the antioxidant pathways. The catalase activity of plants is known to be sensitive to oxidative stress when a lack of iron (or sometimes magnesium) occurs (Esfandiari et al., 2010; Iturbe-Ormaetxe et al., 1995; Tewari et al., 2005) as this protein needs an iron ion in its constitutive heme (Arménia Carrondo et al., 2007). A competitive effect between the excess of copper and the other elements during the adsorption/absorption (Bernal et al., 2007) could lead to a disturbance during the catalase synthesis (Delmail, 2011).

Despite of their extremely toxic nature, the reactive oxygen species are also implied in cascades of signalization which induce the expression and the regulation of many genes. These genes could be involved in the defense mechanisms, like the phytochelatine synthase which allows the synthesis of heavy-metal binding peptides, the phytochelatins. These compounds play important roles in the detoxification of toxic heavy metals and the regulation of intracellular concentrations of essential metals in plants (Hirata et al., 2005). The primary structure of phytochelatins generally have the form (γ-glutamate-cysteine)$_n$-glycine and these peptides could form complexes with heavy metals such as cadmium (Fig. 6), copper, zinc, mercury, silver and arsenic, which are stored as inactive in the cell vacuoles. The expression of phytochelatin synthase in *Populus tremula x tremuloides* cv. Etrepole transgenic lines expressing the wheat phytochelatin synthase TaPCS1 is stimulated by the presence of heavy metal and this protein aimed at increasing metal tolerance and metal accumulation through overproduction of phytochelatins (Couselo et al., 2010).

The reactive oxygen species are also implied in the apoptosis (or programmed cell death) of plants as regulating agents (Dat et al., 2003; Van Breusegem et al., 2006). This event occurs during all the life of organisms and selected cells or organs are eliminated from the living parts through senescence or abscission to maintain the optimal development of the organisms. During the growth, the apoptosis is involved in several phenomena like the triggering of the aleurone cells to release amylase during the germination of caryopsis, the differentiation of xylem and phloem elements, the development of the root cap and the abscission of leaves (Parent et al., 2008). But plants may use this apoptosis to adapt and to resist towards environmental stress like pathogens. For example, the infection of *Nicotiana obtusifolia* by the downy mildew pathogen *Peronospora tabacina* resulted in a compatible interaction, in which *P. tabacina* penetrated and colonized host leaf tissue (Heist et al., 2004).

This interaction becomes incompatible several days later and it leads to an oxidative burst, with the appearance of necrotic lesions due to reactive oxygen species, which isolates the pathogen from the living parts. This conducts to the inhibition of the pathogen growth. These necrotic lesions are due to hypersensitive cell death in the host and the resistance phenotype was due to the action of a gene known to confer a hypersensitive response, Rpt1 (Heist et al., 2004).

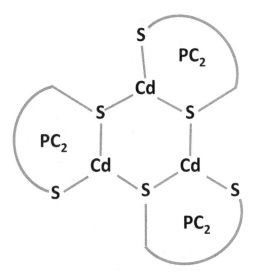

Fig. 6. Schematic structure and organization of phytochelatins implied in the sequestration mechanism of cadmium (Cd) through thiol function (SH) and constituted each of 2 γ-glutamylcysteine parts (PC2) (based on Delmail (2011)).

4. Senescence and abscission

Leaf senescence is a highly regulated process particularly well studied in crop plants and *Arabidopsis* (Balazadeh et al., 2008). Nowadays it is conspicuous that environmental stresses can induce precocious senescence (Balazadeh et al., 2008) as hypothesized since 1997 by Ouzounidou et al. during the observation of the effect of cadmium on wheat; but the effect of heavy metal ions on this phenomenon is still poorly documented. However, it was demonstrated that protein functioning as metal chelator like metallothionein may be needed to protect normal cell functions from the toxic effects of metal ions released during senescence. In that sense, metallothioneins may be involved in chaperoning released metal ions to avoid metal toxicity or metal induced-oxidative stress in plant cell during the senescence process. Guo et al. (2003) indicated that all the *Arabidopsis* metallothionein genes expressed in vegetative tissues were upregulated in senescing leaves thus protecting cells from metal ions toxicity during senescence. A similar observation of the implication of some metallothioneins in leaf senescence and in heavy metal stress was done in barley by Heise et al. (2007).

Another important molecule involved during senescence is the yellow stripe-like transporter family (YSL). Curie et al. (2009) indicated that five out the eight Arabidopsis YSL genes are most strongly expressed in senescent leaves. Indeed, the expression of AtYSL1 and

AtYSL3 is increased during senescence and although the leave of the double ysl1ysl3 mutant loses only 10% of copper content between the 4th and the 5th week of growth, the wild type *Arabidopsis* loses almost 60%. More recently Xiao & Chye (2011) evidenced new roles for acyl-CoA-binding proteins (ACPBs). Indeed, in *Arabidopsis* the expression of AtACPB3 was upregulated during senescence and AtACPB3-KOs *Arabidopsis* displayed delayed leaf senescence whereas AtACPB3-overexpressors *Arabidopsis* present an accelerated leaf senescence phenotype. On the other hand these authors indicated that *Arabidopsis* AtACPB2-overexpressors were more tolerant to cadmium in the growing media.

Among the different mechanisms adopted by plants to cope with metallic stress (phenological escape, exclusion, amelioration and tolerance), the amelioration one implies that the ion must be removed from the circulation or tolerated within the cytoplasm. These amelioration processes include excretion either actively - through glands on aerial part or by roots - or passively by accumulation in old leaves followed by abscission (Adams & Lamoureux, 2005). The simplest form of excretion is the loss of an organ which has accumulated the toxic compounds. This is generally true for the old leaves that present higher content of toxics than the young leaves and buds. For example, Yasar et al. (2006) noticed that the toxic sodium ion was stored in old leaves of the salt-tolerant Gevas Sirik 57 (GS57) green bean genotype acting as a protection mechanism from the detrimental effect of sodium for young leaves. In the same way, Szarek-Lukaszewska et al. (2004) indicated that an *Armeria maritima* population from metalliferous soil directed to the oldest leaves a part of the metal transported to aboveground plant organs. For these authors the ability to accumulate metals in withering leaves characterizes plants growing under strong environmental pressure from metal contamination. Detoxification mechanism by leaf fall was a strategy previously suggested by Dahmani-Mueller et al. (2000) in *Armeria maritima spp. halleri* where metal content (cadmium, copper, lead and zinc) in ageing leaves (brown leaves) were 3-8 times higher than in green leaves. A similar observation was done by Monni et al. (2001) on a shrub (*Empetrum nigrum*) which accumulates metals (cadmium, copper, iron, lead, nickel and zinc) in older tissues, mainly leaves and bark, by both accumulation and surface contamination. For tree species, Pahalawattaarachchi et al. (2009) shown that in *Rhizophora mucronata* chromium, cadmium and lead were accumulated in leaves before abscission and thus eliminated. A major disadvantage of the excretion strategy for plant is that they are stationary so the excreted substance will remain in the root zone and may eventually lead to a build-up of the xenobiotic (Adams & Lamoureux, 2005).

Only few data were available on aquatic macrophyte, a case where this major disadvantage did not apply. For example in *Spirodela polyrrhiza*, the excess of iron and copper induces plant necrosis, colony disintegration and root abscission (Xing et al., 2010). It should be noted that in another aquatic macrophyte, *Lemna minor*, the frond abscission could be used to test water toxicity induced by metal and other compounds (Henke et al., 2011). Our previous data (Delmail et al., 2011d) suggest that as in terrestrial plants a similar excretion strategy could occur in aquatic plants. Indeed, *Myriophyllum alterniflorum* old leaves are much more affected by heavy-metal pollution than younger ones. Previous study of Jana & Chouduri (1982) on three submerged aquatic macrophytes (*Potamogeton pectinatus, Vallisneria spiralis* and *Hydrilla verticillata*) demonstrated that all the heavy metals tested (cadmium, copper, lead and mercury) hastened the senescence process. These authors evidenced the role of the plant growth regulator kinetin in the reduction of the senescence induced by heavy metals.

The role of plant growth regulator in senescence and in heavy-metal resistance is quite complex but cytokinins for their senescence delaying action (for a review see Werner & Schmulling, 2009) and brassinosteroids for their role in responses to various environmental stress (for a review see Bajguz & Hayat, 2009) appeared as major candidates for further studies to understand the heavy-metal induced senescence processes. Indeed, Arora et al. 2010) demonstrated that the brassinosteroid 24-epibrassinolide present stress-ameliorative properties in *Brassica juncea* plant during chromium stress as an improved growth and antioxidant enzymes activities. Similar conclusion was highlighted by Anuradha & Rao 2007) on *Raphanus sativus* plant were brassinosteroids supplementation alleviated the toxic effect of cadmium. More recently, Bajguz (2011) noted on *Chlorella vulgaris* that the brassinosteroid application to the culture prevents chlorophyll, sugar and protein loss and increases phytochelatin synthesis during heavy metal stress (cadmium, copper and lead). These reactions call to mind the delayed senescence process observed previously in aquatic macrophytes when treated with cytokinins during a heavy-metal stress. In the same way, brassinosteroid treatment improves sunflower (genotype 2603) and turnip (var. rave du Limousin) resistance to cadmium stress in terms of photosynthesis activities (Figs. 7 and 8, Delmail et al. unpublished data).

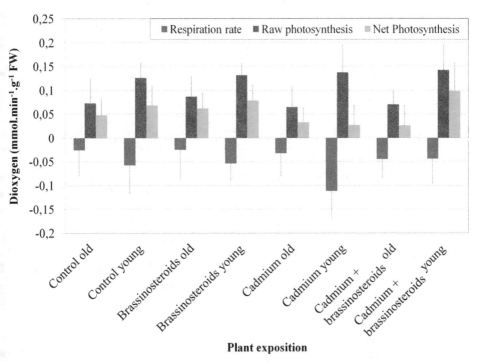

Fig. 7. Photosynthetic activities and respiration rate of one-month old sunflower specimens after 48h of cadmium exposure (1 mM) combined or not with 3 µM 24-epibrassinolid (Delmail et al. unpublished data). FW, fresh weight.

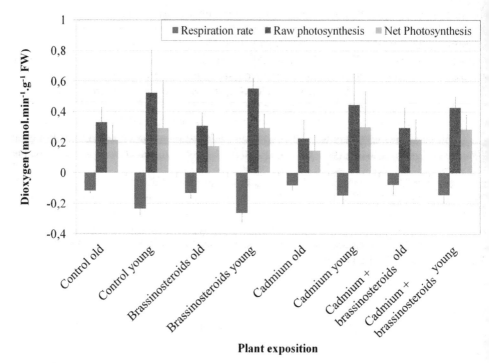

Plant exposition

Fig. 8. Photosynthetic activities and respiration rate of one-month old turnip specimens after 72h of cadmium exposure (1 mM) combined or not with 3 µM 24-epibrassinolid (Delmail et al. unpublished data). FW, fresh weight.

Moreover, the effect of brassinosteroids on antioxidant enzymatic activities during cadmium stress could be similar in young and old leaves as shown in Fig. 9 for a decrease in catalase activity of sunflower plants treated with phytohormons during a heavy-metal stress. On the other hand, as demonstrated in turnip (Fig. 10) a differential effect between old and young leaves could appear with an increase in superoxide-dismutase activity in young leaves and a decrease in old leaves (Delmail et al. unpublished data). In these two plants, brassinosteroid application clearly has a protective effect on the raw photosynthesis activity, probably indicating a delayed heavy-metal senescence. These protective actions probably also occur on the enzymatic antioxidant system even if the complexity of the involved cascade reactions lead to a more unclear landscape inducing pattern variations between studied enzymes, age of plant parts and plant species. It appears clearly that much more studies are needed to understand the complex interwoven relationship existing between plant physiology under heavy-metal stress, senescence and plant growth regulators.

Concerning the organic xenobiotics effect on plant senescence even much less data are available. For example, Cape et al. (2003) noted that in *Lotus corniculatus* exposes to a mixture of six volatile organic compounds (acetone, acetonitrile, dichloromethane, ethanol, methyl t-butyl ether and toluene), a premature senescence occurs but in this case a premature senescence refers to advanced timing of seed pot production. Another example

oncerns the die-back symptom of *Phragmites communis* where the premature senescence of
hoot appears to result at least in part from phytotoxin action (acetic, propionic, n- and iso-
utyric and n-caproic acids and sulphide) (Armstrong & Armstrong, 2001).

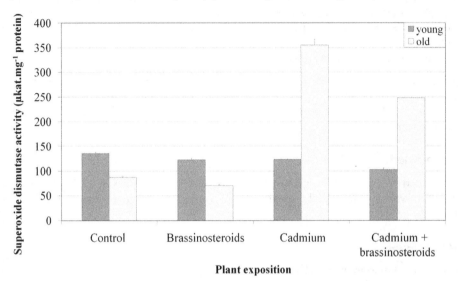

Fig. 9. Superoxide dismutase activity of one-month old sunflower specimens after 48h of
admium exposure (1 mM) combined or not with 3 µM 24-epibrassinolid (Delmail et al.
unpublished data).

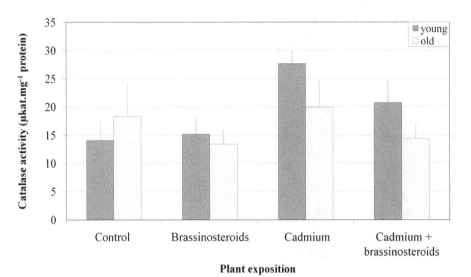

Fig. 10. Catalase activity of one-month old turnip specimens after 72h of cadmium exposure
1 mM) combined or not with 3 µM 24-epibrassinolid (Delmail et al. unpublished data).

5. Conclusion

Senescence implies a succession of physiological events integrated with developmental program which lead to the loss of several organs from the plant. This biological process constitutes an integral part of the normal plant developmental cycle which can be observed at different organization levels (cell, tissue and organ). The senescence is the final event in the life of many plant tissues and it is a highly regulated process that involves structural, biochemical and molecular changes.

Organic and inorganic xenobiotics could hasten the senescence processes and they may lead to a premature death of the plants. At the opposite, the senescence occurring in the plant organs could isolate the stressors and/or eliminate the toxics from the living parts through induced abscission.

6. Acknowledgment

This research was partially supported by the University of Limoges, the GRESE (Research Group on Water, Soil and the Environment) EA 4330, and the Conseil Régional du Limousin.

7. References

Adams, P.W., Lamoureux, S. (2005). Resistance to toxicity, In: *A literature review of the use of native Northern plants for the re-vegetation of Arctic mine tailings and mine waste*, 31.08.2011, Available from
  http://www.enr.gov.nt.ca/_live/documents/content/WKSS_Northern_Plants_Re-vegetation-2005.pdf

Alscher, R.G., Erturk, N., Heath, L.S. (2002). Role of superoxide dismutases (SODs) in controlling oxidative stress in plants. *Journal of Experimental Botany*, Vol.53, No.372, (May 2002), pp. 1331-1341, ISSN 0022-0957

Anuradha, A., Rao, S.S.R. (2007). The effect of brassinosteroids on radish (*Raphanus sativus* L.) seedlings growing under cadmium stress. *Plant, Soil and Environment*, Vol.53, No.11, (November 2007), pp. 465-472, ISSN 1214-1178

Arménia Carrondo, M., Bento, I., Matias, P.M., Lindley, P.F. (2007). Crystallographic evidence for dioxygen interactions with iron proteins. *Journal of Biological Inorganic Chemistry*, Vol.12,No.4, (May 2007), pp. 429-442, ISSN 0949-8257

Armstrong, J., Armstrong, W. (2001). An overview of the effects of phytotoxins on *Phragmites australis* in relation to die-back. *Aquatic Botany*, Vol.69, No.2-4, (April 2001), pp. 251-268, ISSN 0304-3770

Arora, P., Bhardwaj, R., Kanwar, M.K. (2010). 24-epibrassinolide regulated diminution of Cr metal toxicity in *Brassica juncea* L. plants. *Brazilian Journal of Plant Physiology*, Vol.22, No.3, (September 2010), pp. 159-165, ISSN 1677-0420

Bajguz, A. (2011). Suppression of *Chlorella vulgaris* growth by cadmium, lead, and copper stress and its restoration by endogenous brassinolide. *Archives of Environmental Contamination and Toxicology*, Vol.60, No.3, (April 2011), pp. 406-416, ISSN 0090-4341

Bajguz, A., Hayat, S. (2009). Effects of brassinosteroids on the plant responses to environmental stresses. *Plant Physiology and Biochemistry*, Vol.47, No.1, (January 2009), pp. 1-8, ISSN 0981-9428

Balazadeh, S., Riaño-Pachón, D.M., Mueller-Roeber, B. (2008). Transcription factors regulating leaf senescence in Arabidopsis thaliana. *Plant Biology*, Vol.10, No.SUPPL.1, (September 2008), pp. 63-75, ISSN, 1435-8603

Bernal, M., Cases, R., Picorel, R., Yruela, I. (2007). Foliar and root Cu supply affect differently Fe-and Zn-uptake and photosynthetic activity in soybean plants. *Environmental and Experimental Botany*, Vol.60, No., (2), pp. 145-150, ISSN 0098-8472

Bienert, G.P., Schjoerring, J.K., Jahn, T.P. (2006). Membrane transport of hydrogen peroxide. *Biochimica et Biophysica Acta – Biomembranes*, Vol.1758, No.8, (August 2006), pp. 994-1003, ISSN 0005-2736

Bienert, G.P., Møller, A.L.B., Kristiansen, K.A., Schulz, A., Møller, I.M., Schjoerring, J.K., Jahn, T.P. (2007). Specific aquaporins facilitate the diffusion of hydrogen peroxide across membranes. *The Journal of Biological Chemistry*, Vol.282, No.2, (January 2007), pp. 1183-1192, ISSN 0021-9258

Cape, J.N., Leith, I.D., Binnie, J., Content, J., Donkin, M., Skewes, M., Price, D.N., Brown, A.R., Sharpe, A.D. (2003). Effects of VOCs on herbaceous plants in an open-top chamber experiment. *Environmental Pollution*, Vol.124, No.2, (July 2003), pp. 341-353, ISSN 0269-7491

Couselo, J.L., Navarro-Aviñó, J., Ballester, A. (2010). Expression of the phytochelatin synthase TaPCS1 in transgenic aspen, insight into the problems and qualities in phytoremediation of Pb. *International Journal of Phytoremediation*, Vol.12, No.4, (April 2010), pp. 358-370, ISSN 1522-6514

Curie, C., Cassin, G., Couch, D., Divol, F., Higuchi, K. , Le Jean, M. , Misson, J. , Schikora, A. , Czernic, P., Mari, S. (2009). Metal movement within the plant: Contribution of nicotianamine and yellow stripe 1-like transporters. *Annals of Botany*, Vol.103, No.1, (January 2009), pp. 1-11, ISSN 0305-7364

Dahmani-Muller, H., Van Oort, F., Gélie, B., Balabane, M. (2000). Strategies of heavy metal uptake by three plant species growing near a metal smelter. *Environmental Pollution*, Vol.109, No.2, (August 2000), pp. 231-238, ISSN 0269-7491

Dat, J.F., Pellinen, R., Beeckman, T., Van De Cotte, B., Langebartels, C., Kangasjarvi, J., Inze, D., Van Breusegem, F. (2003). Changes in hydrogen peroxide homeostasis trigger an active cell death process in tobacco. *The Plant Journal*, Vol.33, No.4, (February 2003), pp. 621-632, ISSN 0960-7412

Delmail, D. (2007). Ecologie florale bretonne. *La Garance Voyageuse*, Vol.79, No.1, (September 2007), pp. 14-17, ISSN 0988-3444

Delmail, D. (2011). *Contribution de Myriophyllum alterniflorum et de son périphyton à la biosurveillance de la qualité des eaux face aux métaux lourds*, Université de Limoges, Limoges, France

Delmail, D., Labrousse, P., Buzier, R., Botineau, M. (2009). Importance of *Myriophyllum alterniflorum* D.C., an aquatic macrophyte, in biomonitoring of trace metal pollution in running freshwater, *Proceedings of EMEC10 10th European Meeting on Environmental Chemistry*, pp. 66, Limoges, France, December 2-5, 2009

Delmail, D., Labrousse, P., Hourdin, P., Botineau, M. (2010). Use of *Myriophyllum alterniflorum* (Haloragaceae) for restoration of heavy-metal-polluted freshwater environments: preliminary results, *Proceedings of 7th SER European Conference on Ecological restoration and sustainable development*, pp. 103, Avignon, France, August 23-27, 2010

Delmail, D., Labrousse, P., Crassous, P., Hourdin, P., Guri, M., Botineau, M. (2011a). Simulating the dynamics of epiphytic diatom metacommunity in stream environments contaminated with heavy metals, *Proceedings of ECEM 2011 7th European Conference on Ecological hierarchy from the genes to the biosphere*, pp. 112, Riva del Garda, Italy, May 30-June 2, 2011

Delmail, D., Labrousse, P., Hourdin, P., Botineau, M. (2011b). Evidence of copper impact on freshwater environments using *Myriophyllum alterniflorum*: restoration, biomonitoring and management, *Proceedings of IAVS 2011 54th International Symposium on Vegetation in and around water: patterns, processes and threats*, pp. 157, ISBN 978-2-9539515-1-6, Lyon, France, June 20-24, 2011

Delmail, D., Labrousse, P., Hourdin, P., Larcher, L., Moesch, C., Botineau, M. (2011c). Differential responses of *Myriophyllum alterniflorum* DC (Haloragaceae) organs to copper: physiological and developmental approaches. *Hydrobiologia*, Vol.664, No.1, (April 2011), pp. 95-105, ISSN 0018-8158

Delmail, D., Labrousse, P., Hourdin, P., Larcher, L., Moesch, C., Botineau, M. (2011d). Physiological, anatomical and phenotypical effects of a cadmium stress in different-aged chlorophyllian organs of *Myriophyllum alterniflorum* DC (Haloragaceae). *Environmental and Experimental Botany*, Vol.72, No.2, (September 2011), pp. 174-181, ISSN 0098-8472

Edreva, A. (2005). Generation and scavenging of reactive oxygen species in chloroplasts: a submolecular approach. *Agriculture, Ecosystems and Environment*, Vol.106, No.2-3, (April 2005), pp. 119-133, ISSN 0167-8809

Esfandiari, E., Shokrpour, M., Alavi-Kia, S. (2010). Effect of Mg deficiency on antioxydant enzymes activities and lipid peroxidation. *Journal of Agricultal Science*, Vol.2, No.3, (September 2010), pp. 131-136, ISSN 1916-9752

Fornazier, R.F., Ferreira, R.R., Vitória, A.P., Molina, S.M.G., Lea, P.J., Azevedo, R.A. (2002). Effects of cadmium on antioxidant enzyme activities in sugar cane. *Biologia Plantarum*, Vol.45, No.1, (January 2002), pp. 91-97, ISSN 0006-3134

Gill, S.S., Tuteja, N. (2010). Reactive oxygen species and antioxidant machinery in abiotic stress tolerance in crop plants. *Plant Physiology and Biochemistry*, Vol.48, No.12, (December 2010), pp. 909-930, ISSN 0981-9428

Guo, W.-J., Bundithya, W., Goldsbrough, P.B. (2003). Characterization of the *Arabidopsis* metallothionein gene family: Tissue-specific expression and induction during senescence and in response to copper. *New Phytologist*, Vol.159, No.2, (August 2003), pp. 369-381, ISSN 0028-646X

Heise, J., Krejci, S., Miersch, J., Krauss, G.-J., Humbeck, K. (2007). Gene expression of metallothioneins in barley during senescence and heavy metal treatment. *Crop Science*, Vol.47, No.3, (May 2007), pp. 1111-1118, ISSN 0011-183X

Heist, E.P., Zaitlin, D., Funnell, D.L., Nesmith, W.C., Schardl, C.L. (2004). Necrotic lesion resistance induced by *Peronospora tabacina* on leaves of *Nicotiana obtusifolia*. *Phytopathology*, Vol.94, No.11, (November 2004), pp. 1178-1188, ISSN 0031-949X

Ienke, R., Eberius, M., Appenroth, K.-J. (2011). Induction of frond abscission by metals and other toxic compounds in *Lemna minor*. *Aquatic Toxicology*, Vol.101, No.1, (January 2011), pp. 261-265, ISSN 0166-445X

Iinsinger, P., Schneider, A., Dufey J.E. (2005). Le sol : ressource en nutriments et biodisponibilité, In : *Sols et Environnement*, M.-C. Girard, C. Walter, J.-C. Rémy, J. Berthelin, J.-L. Morel (Eds.), 285-305, Dunod, ISBN 978-2-1005-1695-7, Paris, France

Iirata, K., Tsuji, N., Miyamoto, K. (2005). Biosynthetic regulation of phytochelatins, heavy-metal-binding peptides. *Journal of Bioscience and Bioengineering*, Vol.100, No.6, (December 2005), pp. 593-599, ISSN 1389-1723

turbe-Ormaetxe, I., Moran, J.F., Arrese-Igor, C., Gogorcena, Y., Klucas, R.V., Becana, M. (1995). Activated oxygen and antioxidant defences in iron-deficient pea plants. *Plant, Cell and Environment*, Vol.18, No.4, (April 1995), pp. 421-429, ISSN 0140-7791

ana, S., Choudhuri, M.A. (1982). Senescence in submerged aquatic angiosperms: Effects of heavy metals. *New Phytologist*, Vol.90, No.3, (), pp. 477-484, ISSN 0028-646X

Kabata-Pendias, A., Pendias, H. (2000). *Trace elements in soils and plants*, CRC Press, ISBN 978-0-8493-1575-6, Boca Raton, Florida, USA

Kruger, N.J., von Schaewen, A. (2003). The oxidative pentose phosphate pathway: structure and organisation. *Current Opinion in Plant Biology*, Vol.6, No.3, (June 2003), pp. 236-246, ISSN 1369-5266

Lagadic, L. , Caquet, T., Amiard J.-C. (1997). Biomarqueurs en écotoxicologie : principes et définitions, In: *Biomarqueurs en écotoxicologie*, L. Lagadic, T. Caquet, J.C. Amiard, (Eds.), 1-10, Masson Press, ISBN 2-225-83053-3, Paris, France

Li, Y., Trush, M.A., Yager, J.D. (1994). DNA damage caused by reactive oxygen species originating from a copper-dependent oxidation of the 2-hydroxy catechol of estradiol. *Carcinogenesis*, Vol.15, No.7, (July 1994), pp. 1421-1427, ISSN 0143-3334

Monni, S., Uhlig, C., Junttila, O., Hansen, E., Hynynen, J. (2001). Chemical composition and ecophysiological responses of *Empetrum nigrum* to aboveground element application. *Environmental Pollution*, Vol.112, No.3, (May 2001), pp. 417-426, ISSN 0269-7491

Ouzounidou, G., Moustakas, M., Eleftheriou, E.P. (1997). Physiological and ultrastructural effects of cadmium on wheat (*Triticum aestivum* L.) leaves. *Archives of Environmental Contamination and Toxicology*, Vol.32, No.2, (February 1997), pp. 154-160, ISSN 0090-4341

Pahalawattaarachchi, V., Purushothaman, C.S., Vennila, A. (2009). Metal phytoremédiation potential of Rhizophora mucronata (Lam.). *Indian Journal of Marine Sciences*, Vol.38, No.2, (June 2009), pp. 178-183, ISSN 0379-5136

Parent, C., Capelli, N., Dat, J. (2008). Formes réactives de l'oxygène, stress et mort cellulaire chez les plantes. *Comptes Rendus Biologies*, Vol.331, No.4, (April 2008), pp. 255-261, ISSN 1631-0691

Paramesha, S., Vijay, R., Bekal, M., Kumari, S., Pushpalatha, K.C. (2011). A study on lipid peroxidation and total antioxidant status in diabetes with and without hypertension. *Research Journal of Pharmaceutical, Biological and Chemical Sciences*, Vol.2, No.1, (January-March 2011), pp. 329-334, ISSN 0975-8585

Pereira, G.J.G., Molina, S.M.G., Lea, P.J., Azevedo, R.A. (2002). Activity of antioxidant enzymes in response to cadmium in *Crotalaria juncea*. Plant and Soil, Vol.239, No.1, (January 2002), pp. 123-132, ISSN 0032-079X

Qiu, J., Zhu, K., Zhang, J., Zhang, Z. (2002). Diversity of the microorganisms degrading aromatic hydrocarbons. *Chinese Journal of Applied Ecology*, Vol.13, No.12, (December 2002), pp. 1713-1715, ISSN 1001-9332

Ritter, L., Solomon, K., Sibley, P., Hall, K., Keen, P., Mattu, G., Linton, B. (2002). Sources pathways, and relative risks of contaminants in surface water and groundwater: A perspective prepared for the Walkerton inquiry. *Journal of Toxicology and Environmental Health - Part A*, Vol.65, No.1, (January 2002), pp. 1-142, ISSN 1528-7394

Sacco, M.G., Amicone, L., Catò, EM., Filippini, D., Vezzoni, P., Tripodi, M. (2004). Cell-based assay for the detection of chemically induced cellular stress by immortalized untransformed transgenic hepatocytes. *BMC Biotechnology*, Vol.4, No.5, (March 2004), pp. 1-7, ISSN 1472-6750

Schwab, A.P. (2005). Pollutants: Organic and Inorganic, In: *Encyclopedia of Soil Science*, L Rattan, (Ed.), 1334-1337, CRC Press, ISBN 978-0-8493-3830-4

Schaich, K.M. (2005). Lipid oxidation: theoretical aspects, In: *Bailey's Industrial Oil and Fat Products*, F. Shahidi, (Ed.), 269-355, Wiley J., Sons Publishers, ISBN 978-0-471-38460-1

Szarek-Łukaszewska, G., Słysz, A., Wierzbicka, M. (2004). Response of *Armeria maritima* (Mill.) Willd. to Cd, Zn and Pb. *Acta Biologica Cracoviensia Series Botanica*, Vol.46, No.1, (December 2004), pp. 19-24, ISSN 0001-5296

Tewari, R., Kumar, P., Sharma, P. (2005). Signs of oxidative stress in the chlorotic leaves of iron starved plants. *Plant Science*, Vol.169, No.6, (December 2005), pp. 1037-1045, ISSN 0168-9452

Thompson, J.E., Legge, R.L., Barber, R.F. (1987). The role of free radicals in senescence and wounding. *New Phytologist*, Vol.105, No.3, (March 1987), pp. 317-344, ISSN 0028-646X

Van Breusegem, F., Dat, J.F. (2006). Reactive oxygen species in plant cell death. *Plant Physiology*, Vol.141, No.2, (June 2006), pp. 384-390, ISSN 0032-0889

Werner, T., Schmülling, T. (2009). Cytokinin action in plant development. *Current Opinion in Plant Biology*, Vol.12, No.5, (October 2009), pp. 527-538, ISSN 1369-5266

Xiao, S., Chye, M.-L. (2011). New roles for acyl-CoA-binding proteins (ACBPs) in plant development, stress responses and lipid metabolism. *Progress in Lipid Research*, Vol.50, No.2, (April 2011), pp. 141-151, ISSN 0163-7827

Xing, W., Huang, W., Liu, G. (2010). Effect of excess iron and copper on physiology of aquatic plant *Spirodela polyrrhiza* (L.) schleid. *Environmental Toxicology*, Vol.25, No.2, (April 2010), pp. 103-112, ISSN 1520-4081

Yasar, F., Uzal, O., Tufenkci, S., Yildiz, K. (2006). Ion accumulation in different organs of green bean genotypes grown under salt stress. *Plant, Soil and Environment*, Vol.52, No. 10, (October 2006), pp. 476-480, ISSN 1214-1178

8

The Legume Root Nodule: From Symbiotic Nitrogen Fixation to Senescence

Laurence Dupont, Geneviève Alloing, Olivier Pierre,
Sarra El Msehli, Julie Hopkins, Didier Hérouart and Pierre Frendo
*[1]UMR "Biotic Interactions and Plant Health" INRA 1301-CNRS 6243
University of Nice-Sophia Antipolis, F-06903 Sophia-Antipolis Cedex,
[2]Laboratory of Plant Physiology, Science University, Tunis,*
[1]France
[2]Tunisia

1. Introduction

Biological nitrogen fixation (BNF) is the biological process by which the atmospheric nitrogen (N_2) is converted to ammonia by an enzyme called nitrogenase. It is the major source of the biosphere nitrogen and as such has an important ecological and agronomical role, accounting for 65 % of the nitrogen used in agriculture worldwide. The most important source of fixed nitrogen is the symbiotic association between rhizobia and legumes. The nitrogen fixation is achieved by bacteria inside the cells of *de novo* formed organs, the nodules, which usually develop on roots, and more occasionally on stems. This mutualistic relationship is beneficial for both partners, the plant supplying dicarboxylic acids as a carbon source to bacteria and receiving, in return, ammonium. Legume symbioses have an important role in environment-friendly agriculture. They allow plants to grow on nitrogen poor soils and reduce the need for nitrogen inputs for leguminous crops, and thus soil pollution. Nitrogen-fixing legumes also contribute to nitrogen enrichment of the soil and have been used from Antiquity as crop-rotation species to improve soil fertility. They produce high protein-containing leaves and seeds, and legumes such as soybeans, groundnuts, peas, beans, lentils, alfalfa and clover are a major source of protein for human and animal consumption. Most research concentrates on the two legume-rhizobium model systems *Lotus-Mesorhizobium loti* and *Medicago-Sinorhizobium meliloti*, with another focus on the economically-important *Glycine max* (soybean) *-Bradyrhizobium japonicum* association. The legume genetic models *Medicago truncatula* and *Lotus japonicus* have a small genome size of *ca.* 450 Mbp while *Glycine max* has a genome size of 1,115 Mbp, and all are currently targets of large-scale *genome sequencing* projects (He et al., 2009; Sato et al., 2008; Schmutz et al., 2010). The complete genome sequence of their bacterial partners has been established (Galibert et al., 2001; Kaneko et al., 2000; Kaneko et al., 2002; Schneiker-Bekel et al., 2011).

1.1 Early interaction and nodule development

Symbiotic interaction begins with the infection process, which is initiated by a reciprocal exchange of signals between plant and the compatible bacteria. Aromatic compounds -

mostly flavonoids - are secreted by the plants into the rhizosphere and activate the bacterial NodD proteins that are members of the LysR family of transcriptional activators, which in turn induce the expression of the *nod* genes (Long, 2001). This results in the secretion by the bacteria of lipo-chitin oligosaccharide molecules called Nod factors, which are recognized by epidermal cells via specific receptor kinases containing extracellular LysM domains. The spectrum of flavonoids exuded by a legume, as well as the strain-specific chemical structures of the Nod factors, are primary determinants of host specificity (Broughton et al., 2000). Additional bacterial components such as exopolysaccharides, type III and type IV secretion systems are also required for an effective infection (Perret et al., 2000; Saeki, 2011).

Nod factor perception initiates a complex signalling pathway essential for bacterial invasion of the host plant and formation of the nodule. Nod factor signal transduction requires a calcium signalling pathway, which includes the activation of a calcium and calmodulin dependent protein kinase in response to nuclear calcium oscillations. The ensuing induction of gene expression results in rearrangements of the root hair cytoskeleton and initiation of bacterial infection at the epidermis. The root hairs curl and trap the rhizobia which enter the root hair through tubular structures called infection threads. Simultaneously Nod factors induce root cortex cells dedifferentiation and division, leading to the formation of nodule primordia which then differentiate into N_2-fixing nodules (Crespi & Frugier, 2008; Oldroyd et al. 2011). The growing infection threads traverse the root epidermis and cortex, penetrate primordial cells, and then invading bacteria are released into the host cells by an endocytosis-like mechanism (Ivanov et al., 2010). Each bacterium is surrounded by a plant cell membrane, the peribacteroid membrane (PBM), the whole forming an organelle-like structure called the symbiosome where bacteria differentiate into nitrogen-fixing bacteroids. These symbiosomes ultimately completely fill the cytoplasm of infected cells. As the bacteria differentiate, infected cells undergo enlargement coupled to repeated endoreduplication cycles - genomic DNA replication without mitosis or cytokinesis - and become large polyploid cells housing thousands of bacteroids (Jones et al., 2007; Kondorosi et al., 2000). Mature nodules actively fix nitrogen until they enter senescence upon aging or stress.

Nodule organogenesis is accompanied by major changes in plant gene expression. Several hundred of genes were found to be strongly and specifically up- or -down regulated during the nodulation process (Benedito et al., 2008; El Yahyaoui et al., 2004). In *M. truncatula*, two distinct waves of gene expression reprogramming accompany the differentiation of both the plant infected cell and bacteroids (Maunoury et al., 2010). Genes exclusively expressed or strongly up-regulated in nodules have been termed "nodulins". The early nodulins are involved in signal transduction and nodule development and the late nodulins are induced when N_2 fixation begins. Different expression profiling tools relying on genome and high-throughput EST-sequencing have been developed to identify nodulin genes on a large scale (Kuster et al., 2007; Schauser et al., 2008).

Nodules can be classified into two main groups according to their mode of development (Franssen et al., 1992; Maunoury et al., 2008) (Figure 1). Legumes such as *Phaseolus vulgaris* (bean), *Lotus japonicus* or *Glycine max* (soybean) form determinate nodules that have no permanent meristem and adopt a globular shape. The mature nodules contain a homogenous central tissue composed of infected cells fully packed with nitrogen-fixing bacteroids and some uninfected cells. Senescence in these nodules occurs radially, beginning at the center and extending to the periphery. Decaying nodules release bacteroids most of which are able to revert to a free-living lifestyle. Conversely, legumes such as *Medicago truncatula, Pisum sativum* (pea) or *Trifolium* (clover) form indeterminate nodules that possess a permanent meristem and elongate, to become cylindrical. In mature nodule of this type, several histological zones of consecutive developmental states can be distinguished (Vasse et al., 1990). The apical meristem, free of bacteria, is the zone I. Zone II is the infection zone in which post-mitotic cells enter the nodule differentiation programme and where infection threads penetrate the plant cells and release rhizobia. In zone III, the bacteroids are able to fix N_2. A root proximal senescence zone (zone IV) can be observed in older nodules, where the bacteroids, together with the plant cells, are degraded. Upon aging this zone gradually extends to reach the apical part and the nodule degenerates. Proximal to the zone IV is a region (zone V) containing undifferentiated bacteria, which appear to proliferate in the decaying plant tissue (Timmers et al., 2000). In contrast to bacteroids housed in determinate nodules, those from undeterminate nodules have lost their capacity to reproduce. Thus at the end of symbiosis, essentially bacteria that are released from infection threads can return to a free-living lifestyle and recolonize the rhizosphere (Mergaert et al., 2006).

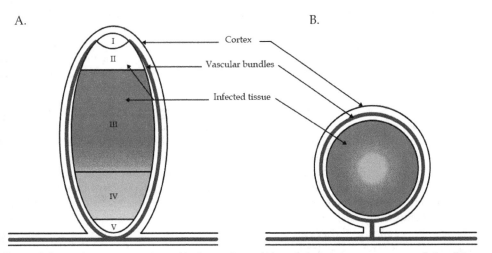

Fig. 1. Schematic representation of indeterminate (A) and determinate mature nodules (B) I, meristemic zone ; II, infection zone ; III, nitrogen fixing zone ; IV, senescence zone ; V, saprophytic zone.

Nitrogen fixing bacteroids in determinate and indeterminate nodules originate from distinct differentiation processes. Bacteroids in legume species forming determinate type nodules present the same cell size, genomic DNA content and reproductive capacity as the free-living bacteria. Conversely, differentiation of bacteroids in indeterminate nodules is linked to drastic morphological and cytological changes, such as cell elongation coupled to genome amplification, membrane permeabilisation and loss of reproductive capacity (Mergaert et al., 2006). This terminal differentiation is mediated by plant-host factors identified as the nodule-specific cysteine-rich (NCR) peptides (Kereszt et al., 2011; Van de Velde et al., 2010). In *M. truncatula* the NCR gene family encodes more than 300 different peptides, which resemble defensine-type antimicrobial peptides (AMPs) (Mergaert et al., 2003).

1.2 Nodule functioning

Bacteria that have completed the bacteroid differentiation program express the enzymes of the nitrogenase complex and begin to fix nitrogen. The reduction by nitrogenase of 1 molecule of N_2 to 2 molecules of NH_4^+ requires 16 molecules of ATP and 8 electrons (Jones et al., 2007). Thus, bacteroids require high rate flux of O_2 to enable high rates of ATP synthesis, but this must be achieved whilst maintaining a very low concentration of free O_2 to avoid inactivation of O_2-labile nitrogenase. These conditions exist due to the presence of an O_2 diffusion barrier and the synthesis of nodule-specific leghemoglobins, which accumulate to millimolar concentrations in the cytoplasm of infected cells prior to nitrogen fixation and buffer the free O_2 concentration at around 7-11 nM, while maintaining high O_2 flux for respiration (Appleby, 1984; Downie, 2005; Ott et al., 2005). The unique low-O_2 environment provided for the bacteroid is a key signal in bacteroid metabolism, inducing a regulatory cascade controlling gene expression of the nitrogenase complex and the microaerobic respiratory enzymes of the bacteroid. The O_2-sensing two-component regulatory system FixL-FixJ activates the transcription of the two intermediate regulators *nifA* and *fixK* genes, which induce the expression of *nif* and *fix* genes involved in nitrogen fixation and respiration (Reyrat et al., 1993). More generally, bacteroid differentiation is accompanied by a global change in gene expression compared with free-living bacteria. There is down-regulation of many genes such as most housekeeping genes and genes involved in synthesis of membrane proteins and peptidoglycan in favour of symbiosis specific processes (Becker et al., 2004; Bobik et al., 2006; Capela et al., 2006; Karunakaran et al., 2009; Pessi et al., 2007).

The reduction of N_2 to ammonium is accompanied, in bacteroids, by the switching-off of ammonium assimilation into amino acids. Ammonium is secreted to the plant cytosol, for assimilation into the amides glutamine and asparagine or into ureides. In return, the plant provides carbon and energy sources to bacteroids in the form of dicarboxylic acids, particularly malate and succinate, which are produced from sucrose via sucrose synthase and glycolytic enzymes. Their metabolization by the TCA cycle provides bacteroids with reducing equivalents, ATP and metabolites for amino acid synthesis and other biosynthetic pathways (White et al., 2007; Lodwig & Poole, 2003). Pea bacteroids also depend on plant for branched-chain amino-acid (LIV) supply, as the bacteroids become symbiotic auxotrophs for these amino-acids (Prell et al., 2009).

The nodule functioning has many peculiarities, involving a plant-microbe crosstalk associated to a metabolism which needs a high energy level under micro-oxic conditions. Nodule development and senescence also have specific features. Whereas multiple review articles have described the early steps of nodule formation and functioning, the rupture of the interaction has not been reviewed recently. In this context, this review focuses on the different characteristics of root nodule senescence.

2. Developmental senescence

For many years, the majority of the research concerning N-fixing symbioses focused on understanding the mechanisms leading to the establishment of this symbiotic relationship, from the invasion of plant cells to the N-fixing bacteroid state. In all nodule types, the N_2-fixation period is optimal between 4 and 5 weeks after infection. Beyond this period, first eductions of N_2-fixing bacteroid capacity are detectable and a senescence process occurs in he N -fixing nodule zone. This phenomena is related to the onset of pod filling in grain egumes like soybean, pea and common bean (Bethlenfalvay & Phillips, 1977; Lawn & Brun, 977). Thus, the lifespan of the rhizobia-plant symbiotic relationship is relatively short and he disruption of this symbiosis affects the yield of the culture.

2.1 Structural analysis of developmental senescence

The spatial dynamics of the senescence process in nodules is nodule type dependent. The pink N_2-fixing tissues of the zone III become green in color in zone IV due to leghemoglobin breakdown (Lehtovaara & Perttila, 1978). In determinate nodules, histological analyses of cross-sections using this simple visible change revealed that senescence develops radially, starting from the center and gradually spreading toward the outside (Puppo et al., 2005). In contrast, in undeterminate nodules, histological analyses of longitudinal sections of nodules based on pink-to-green color changes or based on the expression pattern of bacteroid genes involved in the N_2-fixing process using promoter-lacZ fusions (i.e. NifH) have led many authors to consider the front of senescence as a planar structure (Puppo et al., 2005). Recently, using toluidine blue staining to discriminate senescent from healthy cells and studying the gene expression pattern of a small family of plant cysteine proteases as early markers of nodule senescence on serial transversal sections of M. truncatula nodules, Pérez Guerra and collaborators proposed a conical organization of the developmental senescence zone: the earliest signs of senescence in a few infected cells in the center of the N-fixing zone occurred similar to determined nodules, and this phenomena progressively extended toward the nodule periphery in subsequent proximal cell layers of the nodule (Pérez Guerra et al., 2010).

Comparison between N_2-fixing and senescent cells in soybean nodule showed a decrease of density of plant cytoplasm, the apparition of vesicles associated with the deterioration of symbiosomes and modifications in organelles like peroxisomes, mitochondria and plastids (Lucas et al., 1998; Puppo et al., 2005). Ultrastructural analysis of M. truncatula mature nodule cells has revealed at least two stages during the developmental senescence of N_2-fixing cells: first, a disintegration of bacteroids and symbiosomes revealed by the presence of numerous membranes in the plant cytoplasm associated with the formation of lytic symbiosome compartments probably involved in reabsorption processes and second, the decay of plant infected cells associated with collapse phenomena and that of the plant uninfected cells (Perez Guerra et al., 2010; Timmers et al., 2000; Van de Velde et al., 2006; Vasse et al., 1990). The fusion of symbiosomes to form lytic compartments resembles vacuole formation. Analysis of the relation between the symbiosome formation and the endocytic pathway showed that the lifespan of bacteria in individual symbiosomes compartments during the N_2-fixing stage is achieved by delaying the acquisition of vacuolar identity such as vacuolar SYP22 and VTI11 SNAREs (Limpens et al., 2009). The acquisition of vacuolar identity by symbiosomes upon senescence likely allows the delivery of newly formed proteases to facilitate nutrient remobilization and a sink-to-source transition. Indeed, nodule senescence is accompanied by increased plant proteolytic activities that might cause large-scale protein degradation in soybean (Malik et al., 1981), French bean (Pladys et al., 1991) and alfalfa (Pladys & Vance, 1993).

2.2 Physiological and biochemical modifications during developmental senescence

Developmental nodule senescence is a complex and programmed process which induces a decrease of N_2-fixing activity and leghemoglobin content, modifications in the nodule redox state components and an increase of proteolytic activity, ultimately leading to the death of infected cells.

Leghemoglobin (Lb), which has a fundamental role in nodule functioning, is an important physiological marker for following the progression of nodule senescence (Figure 2). Lb content progressively decreases with the onset of senescence. This diminution of Lb impacts not only general metabolism of nodule by decreasing the O_2 availability to bacteroids with a low free O_2 content but also by potentially releasing free iron to produce reactive oxygen species (ROS) via the Fenton reaction. Indeed, the auto-oxidation of the active form of Lb, ferro-Lb-O_2 (Lb-Fe^{2+}-O_2), is associated with superoxide anion (O_2^-) production (Fridovich, 1986; Puppo et al., 1981) and the degradation of the heme group of Lb by H_2O_2 likely allows the release of the catalytic Fe which enhances the production of OH· through the Fenton and the Haberweiss reactions (Becana & Klucas, 1992; Puppo & Halliwell, 1988). The importance of Lb in nodule ROS production has been shown in a transgenic *Lotus japonicus* line in which diminution of the Lb content is correlated with diminution of the H_2O_2 production (Gunther et al., 2007).

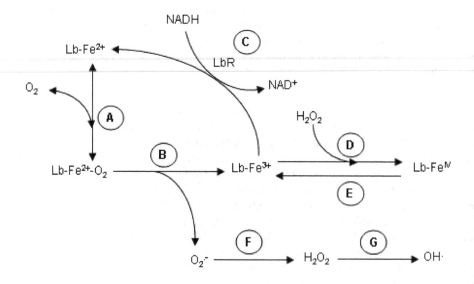

Fig. 2. Metabolic pathways involving Lb and formation of ROS.
A, Reversible oxygenation of Lb-Fe^{2+}; B, Autoxidation of Lb-Fe^{2+}-O_2 to Lb-Fe^{3+} with release of O_2^-; C, Lb-Fe^{3+} reduction by ferric Lb reductase (LbR); D, H_2O_2 reaction with Lb-Fe^{3+} to generate the inactive Lb-Fe^{IV} (ferryl) form; E, Lb-Fe^{IV} reduction to Lb-Fe^{3+} by ascorbate or thiols; other ROS can also be generated from O_2^-, by its dismutation to H_2O_2; F, and by H_2O_2 reduction to OH· through Fenton reaction (G).

Large modifications of the redox balance occur upon natural nodule senescence. Redox balance is defined by the equilibrium between the production of ROS and their degradation by the antioxidant defence system (Apel & Hirt, 2004). Ascorbate (Asc), homoglutathione (hGSH) and glutathione (GSH) are major antioxidants and redox buffers in plant nodule cells (Becana et al., 2010). The regulation of Asc and hGSH biosynthesis has been studied in common bean (*Phaseolus vulgaris*) nodules during aging (Loscos et al., 2008). The expression of five genes of the major Asc biosynthetic pathway was analyzed in nodules, and evidence was found that L-galactono-1,4-lactone dehydrogenase (GalLDH) , the last committed step of the pathway, is post-transcriptionally regulated. Large differences of Asc concentrations and redox states were observed in *P. vulgaris* nodules at different senescence stages suggesting that the lifespan of nodules is in part controlled by endogenous factors like Asc. Biochemical assays on alfalfa dissected nodules revealed that the senescent zone had lower GalLDH activity and ascorbate concentration compared to the infected zone (Matamoros et al., 2006). A strong positive correlation between N_2-fixing activity and nodule Asc and GSH contents was also observed during pea nodule development and senescence (Groten et al., 2005). Peroxiredoxins (Prx) have also been described in N_2-fixing nodules (Groten et al., 2006). Pea nodules contain at least two isoforms of Prx, located potentially in the cytosol (PrxIIB / C) and mitochondria (PrxIIF). The levels of PrxIIB /C declined with nodule senescence, but those of PrxIIF remained unaffected (Groten et al., 2006). The progressive decrease of antioxidant content during pea nodule senescence is not accompanied by an increase in ROS such as O_2^- and H_2O_2 (Groten et al., 2005). In contrast, in aging soybean nodules, an oxidative stress has been detected including an increase of ROS, oxidized hGSH, catalytic Fe and oxidatively modified proteins and DNA bases, but no changes in Asc or tocopherol (Evans et al., 1999). The imbalance in redox state leading to oxidative stress induces the oxidation of lipids and proteins and the degradation of membranes. Lipid peroxidation was found to be elevated in senescent nodules of pigeon pea (*Cajanus cajan*) and bean (Loscos et al., 2008; Swaraj et al., 1995). In senescent soybean nodules, the presence of large amounts of H_2O_2 in the cytoplasmic and apoplastic compartments of the central infected tissue was detected and associated with a widespread expression of a cysteine protease gene (Alesandrini et al., 2003), suggesting a link between oxidative stress and proteolitic activities detected upon nodule senescence.

Various proteases, including those of the acid, serine, aspartic and cysteine types, have been isolated from senescing nodule tissue of soybean, alfalfa, French bean, and pea (Kardailsky & Brewin, 1996; Malik et al., 1981; Pfeiffer et al., 1983; Pladys & Vance, 1993; Pladys & Rigaud, 1985). The induction of cysteine protease genes during nodule senescence has been shown in soybean (Alesandrini et al., 2003), Chinese milk vetch (Naito et al., 2000), pea (Kardailsky & Brewin, 1996) and *M. truncatula* (Fedorova et al., 2002). The general transcriptomic analysis of senescent nodules in *M. truncatula* using cDNA-AFLP (Van de Velde et al., 2006) confirmed the predominant presence of genes encoding representatives of cysteine proteases that are highly homologous to one of the prominent markers of leaf senescence, *Sag12* (Lohman et al., 1994), indicating that these proteinases play an important role in the regulation of developmental nodule senescence. This hypothesis was confirmed in *Astragalus sinicus* since the silencing by RNA interference of the Asnodf32 gene, encoding a nodule-specific cysteine proteinase delayed root nodule senescence with a significant extention of the period of bacteroid active nitrogen fixation. Interestingly, elongated nodules were also observed on Asnodf32-silenced hairy roots (Li et al., 2008).

2.3 Transcriptomic analysis of developmental senescence

The onset of senescence involves the expression of genes whose products are required to carry out senescence-related processes (Gepstein, 2004). In order to isolate genes up- or down-regulated during nodule senescence, several genetic analyses including cDNA libraries and differential screening, mRNA differential display or cDNA-AFLP, have been performed in soybean (Alesandrini et al., 2003; Webb et al., 2008), and *M. truncatula* (Fedorova et al., 2002; Van de Velde et al., 2006). Using a mixture of effective nodules from 7 week-old plants, the first database specific to *M. truncatula* nodule senescence was obtained by isolating 140 000 Expressed Sequence Tags which are available in the J. Craig Venter Institute (http://www.jcvi.org/cms/research/groups/plant-genomics/resources/). To enrich plant material in senescent tissue, recent analyses in *M. truncatula* were performed from cross sections of nodules of 5 and 9 weeks by isolating the zones I, II and III from zones IV and V based on pink-to-green color changes (Van de Velde et al., 2006). This analysis using a modified cDNA-AFLP protocol has resulted in a collection of 508 gene tags that were expressed differentially. Functional clustering of these data has revealed a clear transition from carbon sink to nutriment source for the nodule by up-regulation of genes representative of several different proteases, genes involved in proteasome pathway and degradation of nucleic acids, membrane-derived lipids, and sugars. Moreover, this analysis suggests that three major hormones, ethylene, jasmonic acid and gibberellin, play an important role in nodule senescence (Van de Velde et al., 2006). From a more general point of view, it was been found that a significant overlap exists between genes expressed during leaf senescence in *Arabidopsis thaliana* and nodule senescence in *M. truncatula* (Van de Velde et al., 2006). However, more recent transcriptomic analysis of *M. truncatula* leaf senescence showed that only a minority of common genes are regulated during leaf and nodule senescence (De Michele et al., 2009).

2.4 Hormonal regulation of developmental senescence

Abscisic acid has been proposed to be an important signal in nodule senescence (Puppo et al., 2005), but no direct abscisic acid-responsive genes were present in the cDNA-AFLP dataset from *M. truncatula* nodules (Van de Velde et al., 2006). This analysis revealed that ethylene and jasmonic acid may play a positive role in nodule senescence, just as they do in the senescence of other plant tissues. The positive role of ethylene is illustrated by the up-regulation of ERF transcription factors and ethylene biosynthetic genes, such as S-adenosyl-Met (SAM) synthetase and 1-aminocyclopropane-1-carboxylate oxidase. Involvement of jasmonic acid is suggested by the induction of lipoxygenase genes during different stages of nodule senescence. Moreover, a strong induction of a gene coding for the GA 2-oxidase, that converts active gibberellins to inactive forms (Thomas et al., 1999), was observed in senescent nodule suggesting that gibberellins might repress the senescence process. Finally, the induction of genes encoding a SAM synthase and a spermidine synthethase suggests the involvement of polyamine biosynthetic pathways in nodule senescence. Concerning the potential implication of the two major hormones, auxin and cytokinin, in nodule senescence, only a small amount of data is available. In lupin (*Lupinus albus*), an elevated accumulation of the LaHK1 transcripts, encoding a cytokinin receptor homologue, was detected during nodule developmental senescence suggesting a putative role for this cytokinin receptor homologue in nodule senescence (Coba de la Pena et al., 2008).

3. Stress-induced senescence in legume root nodule

Legume BNF is particularly sensitive and perturbed by environmental stress conditions such as drought, salt stress, defoliation, continuous darkness and cold stress. Adverse environmental conditions affect nodule structure, impair nodule functioning and induce drastic metabolic and molecular modifications leading ultimately to a stress-induced senescence (SIS).

3.1 Stress induced senescence has typical features when compared to developmental senescence

As stated earlier, developmental induced senescence occurs typically in 5 to 11 week old nodules with a slow diminution of BNF during this time period (Evans et al., 1999; Puppo et al., 2005). In contrast, BNF declines drastically and quickly under environmental stress conditions. In less than a week, drought (Gonzalez et al., 1995; Larrainzar et al., 2007; Serraj et al., 1999), salt stress (Soussi et al., 1998; Swaraj and Bishnoi, 1999), dark stress (Matamoros et al., 1999; Gogorcena et al., 1997) and cold stress (van Heerden et al., 2008) decrease dramatically BNF. Thus, SIS is a much faster process than developmental senescence. Moreover, whereas developmental senescence is associated with the establishment of the nodule senescent zone which increases over time, SIS induces the degeneration of the whole nodule in a short time period (Matamoros et al., 1999; Perez Guerra et al., 2010; Vauclare et al., 2010). At the structural level, microscopic analyses also show that developmental senescence and dark stress-induced senescence present different features in *M. truncatula*. Dark-induced senescence leads to the condensation of the bacteroid content whereas the PBM remains intact even though most of the bacteroid content had disappeared (Perez Guerra et al., 2010). In contrast, developmental senescence induces a pronounced vesicle mobilisation in the host cytoplasm which is correlated with the degeneration of the PBM and the mixing of the symbiosome content with the cytoplasm (Perez Guerra et al., 2010; Van de Velde et al., 2006). However, structural analyses of the SIS have not been extensively performed on different nodule types for all the different environmental stress. Thus, it is possible that the different SIS do not develop similarly. Moreover, as for developmental senescence, SIS may process differently in determinate and indeterminate nodules. Indeed, nitrate induced senescence has been shown to induce bacteroid degradation in pea (indeterminate nodule) after two days of treatment in contrast to bean (determinate nodule), in which nitrate has little effect on the shape of bacteroids even after four days of 10mM nitrate treatment (Matamoros et al., 1999).

3.2 Stress induced senescence is characterized by modifications in nodule carbon metabolism and respiration

SIS is characterized by multiple early modifications of nodule physiology (Figure 2). Amongst them, modifications in carbon metabolism play a major role. As stated before, BNF is a highly energetic process which requires a constant energy supply. Modification of nodule sucrose content has been observed during drought stress (Galvez et al., 2005; Gordon et al., 1997), salt stress (Gordon et al., 1997; Lopez et al., 2008; Sanchez et al., 2011; Lopez et al., 2009; Ben Salah et al., 2009), dark stress (Gogorcena et al., 1997; Matamoros et al., 1999; Vauclare et al., 2010) and cold stress (Walsh & Layzell, 1986; van Heerden et al., 2008). However, whereas dark stress induces a diminution of sucrose concentration, drought stress, salt stress and cold stress lead in general to its accumulation in nodules. The diminution of sucrose concentration linked to a shortage in photosynthate feeding result in a deficiency in energy production. In contrast, the accumulation of sucrose during stress suggests that nodule glycolytic enzymes are affected. Sucrose synthase, which is involved in the degradation of sucrose into glucose and fructose in the root nodule, is inhibited during drought stress and salt stress (Ben Salah et al., 2009; Gordon et al., 1997) and malate content, which is one of the preferred substrates for bacteroid respiration (Prell and Poole, 2006), decreases under these two stress conditions (Marino et al., 2007; Galvez et al., 2005; Ben Salah et al., 2009). A fine-tuning of O_2 concentration is also important for nodule functioning, since a good supply of O_2 is determinant for nodule respiration and energy requirement while a low O_2 pressure must be maintained in the nitrogen fixing zone to prevent nitrogenase inhibition. The availability of O_2 through the nodule diffusion barrier and the rate of nodule respiration are thus important parameters of the nodule fitness and they have been shown to be modified during various SIS such as drought stress (Del Castillo et al., 1994; Guerin et al., 1990; Naya et al., 2007; Serraj & Sinclair, 1996; Vessey et al., 1988), salt stress (Bekki et al., 1987; Serraj et al., 1994; Aydi et al., 2004; L'Taief et al., 2007), dark stress (Gogorcena et al., 1997) and cold stress (van Heerden et al., 2008; Kuzma et al., 1995).

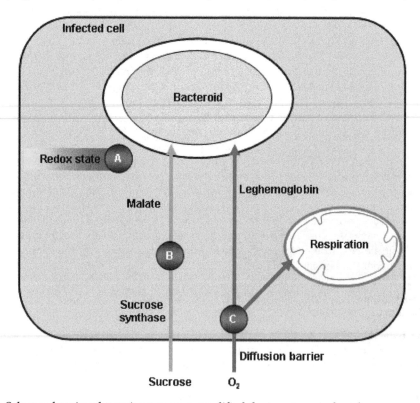

Fig. 3. Scheme showing the major processes modified during stress induced senescence. A, modification of the redox balance; B, alteration of the bacteroid nutrition; C, alteration of O_2 homeostasis.

However, nodule O_2 metabolic modifications are not always similar. Some types of stress decrease nodule permeability to O_2, lowering the O_2 availability to bacteroids, which in turn inhibits nitrogenase activity through a lower nodule respiration rate and lower energetic supply. In this context, a nitrate-nitric oxide respiration process has been identified in nodule which may play a role in the maintenance of energetic status under low oxygen conditions (Horchani et al., 2011). In contrast to stress which reduces oxygen availability, some stress increase nodule O_2 concentration (increased nodule permeability and/or lower respiration rate) which inhibits nitrogenase activity through a direct O_2-induced inactivation.

Leghemoglobin (Lb) is also a general physiological marker of SIS. As mentioned before, Lb has a crucial role in nodule functioning (Ott et al., 2009; Ott et al., 2005). Decrease of Lb content has been shown during drought stress (Gogorcena et al., 1995; Gordon et al., 1997; Guerin et al., 1990), salt stress (Gordon et al., 1997; Mhadhbi et al., 2011) and dark stress (Gogorcena et al., 1997; Matamoros et al., 1999). This diminution of Lb will impact the general metabolism of nodule by decreasing the O_2 availability to bacteroid and by potentially releasing free iron which may be a co-factor of the Fenton reaction to produce reactive oxygen species.

In conclusion, the general production of the high energy level needed for the efficient nitrogen fixation is generally altered at the onset of SIS.

3.3 Stress induced senescence is characterized by modifications in the nodule redox state components

As during developmental senescence, modifications of the redox balance are involved in nodule SIS. As discussed above, the high respiration rates, the important Lb concentration and the release of the catalytic Fe may be major ROS production systems. The accumulation of catalytic Fe has been detected during dark stress (Gogorcena et al., 1997; Becana & Klucas, 1992) and drought stress (Gogorcena et al., 1995) and participates in OH$^{\bullet}$ production during dark stress (Becana & Klucas, 1992). The modification of iron metabolism during SIS is also noticeable through the up regulation of ferritin and metallothionein during drought stress (Clement et al., 2008) and dark stress (Perez Guerra et al., 2010). These proteins sequester free Fe to decrease the Fenton reaction and protect the cellular primary components.

ROS accumulation is also regulated by antioxidant defence which participates in their degradation (Figure 4). Nodule antioxidant defence and the importance of the regulation of the redox balance has been extensively studied in root nodule (for review: Becana et al., 2010; Chang et al., 2009; Marino et al., 2009). Modifications of the antioxidant defence parameters have been used extensively as a marker for nodule SIS. Content and redox state of GSH and Asc, two antioxidant molecules, are modified during drought stress (Gogorcena et al., 1995; Marino et al., 2007), salt stress (Swaraj and Bishnoi, 1999) and dark stress (Matamoros et al., 1999; Gogorcena et al., 1997). Superoxide dismutase and catalase, two enzyme families involved in ROS degradation, are down regulated during dark stress (Matamoros et al., 1999; Gogorcena et al., 1997), salt stress (Jebara et al., 2005) and drought stress (Gogorcena et al., 1995; Rubio et al., 2002). Similarly, enzymes of the Asc-GSH cycle are modulated during nodule SIS (Gogorcena et al., 1995; Jebara et al., 2005; Matamoros et al., 1999; Mhadhbi et al., 2011). Nevertheless, whereas the majority of the reports suggest that SIS is associated with a decrease of the antioxidant defence, other articles have shown

that some elements of the antioxidant defence are stable or even up regulated during stress
This discrepancy may be linked to the stress intensity, the plant adaptation to the treatment
and to the growth conditions which modify the responses of plant to stress.

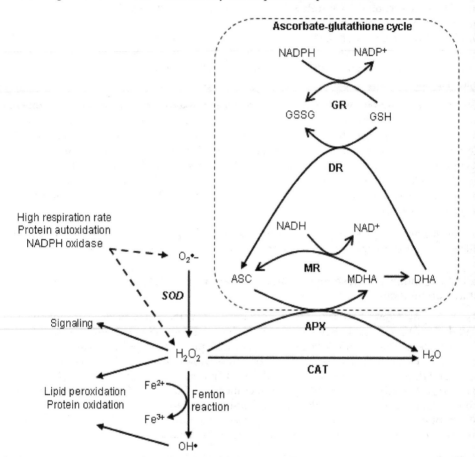

Fig. 4. Scheme showing processes for production and removal of Reactive Oxygen Species.
CAT, catalase; SOD, superoxide dismutase; APX, ascorbate peroxidase; MR,
monodehydroascorbate reductase; DR, dihydroascorbate reductase; GR, glutathione
reductase; GSH, reduced glutathione; GSSG, oxidized glutathione; ASC, reduced ascorbate;
MDHA, monodehydroascorbate; DHA, dehydroascorbate.

The imbalance in redox state leading to oxidative stress is characterized by the oxidation of
major cellular components such as lipids and proteins. One of the major targets is the
membrane. Lipid peroxidation, measured as the reaction of thiobarbituric acid with
malondialdehyde, is significantly attenuated during dark stress (Matamoros et al., 1999;
Gogorcena et al., 1997) in contrast to drought stress and salt stress during which lipid
peroxidation increases (Gogorcena et al., 1995; Mhadhbi et al., 2011). Protein oxidation is

ncreased during dark stress (Matamoros et al., 1999; Gogorcena et al., 1997) and drought stress (Gogorcena et al., 1995). The differential regulation in lipid peroxidation suggests that different degradation mechanisms may partially occur during SIS. This raises the question of SIS regulation and the specificity of the plant response to the different stress.

3.4 Nodule senescence induced by nitrate

One of the specificities of root nodule is the rapid onset of senescence occurring under nitrate treatment. Nitrate concentrations above 2 to 3 mM have strong detrimental effects on the NFS as there is inhibition of several developmental steps ranging from the infection process to the nitrogen fixation in mature nodule (Mortier et al., 2011; Streeter & Wong, 1988). As for other SIS, the inhibition of nitrogenase is correlated with an increase in O_2 diffusion and supply (Escuredo et al., 1996; Matamoros et al., 1999; Minchin et al., 1986; Minchin et al., 1989). Nitrate application also reduces carbon supply from leaves to nodules as measured by plant treatment with ^{11}C and ^{14}C-labelled CO_2 (Fujikake et al., 2003). This reduction in carbon supply is sometimes correlated with the diminution of the sucrose pool (Matamoros et al., 1999) and the down expression of sucrose synthase (Gordon et al., 2002). Nitrate treatment also decreases the antioxidant defence of the nodule with a diminution of the ascorbate pool and of the activities of ascorbate peroxidases and catalases (Escuredo et al., 1996; Matamoros et al., 1999). At the ultrastructural level, the symbiosome membrane seems to be affected by the senescence process before the bacteroid (Matamoros et al., 1999).

Nitrate effect results in both local and systemic regulation of the nodulation process (Jeudy et al., 2009). Systemic regulation has been described in numerous leguminous plants. Indeed, shoot-determined supernodulators with a nitrate-tolerant nodulation process have been described in soybean (Searle et al., 2003), pea (Duc & Messager, 1989), L. japonicus (Krusell et al., 2002) and M. truncatula (Penmetsa et al., 2003). Gene analyses have led to the identification of orthologous Leucine Rich Repeat-Receptor Like Kinases (LRR-RLKs) which play a crucial role in the autoregulation of nodulation. The systemic regulation may occur via the induction of specific CLV3/ESR (CLE) peptides produced after nitrate treatment (Okamoto et al., 2009; Reid et al., 2011). In this context, NOD3 is involved in the production or in the transport of the root signal molecule involved in the systemic regulation (Novak, 2010).

Finally, nitrogen limitation regulates nodule growth and stimulates BNF activity via a LRR-RLKs independent response suggesting that a local nodule adaptation may also be involved in the nitrate regulation of the BNF (Jeudy et al., 2009).

3.5 Molecular modifications occurring during stress induced senescence

The last review analyzing root nodule senescence presented oxidative stress and hormones as potential key players of the senescence process (Puppo et al., 2005). The concentration of abscissic acid (ABA), a hormone involved in plant response to abiotic and biotic stress (Cutler et al., 2010; Raghavendra et al., 2010), is strongly increased in soybean nodules under drought stress (Clement et al., 2008). The five-fold accumulation of ABA in stressed nodules compared to stressed roots shows that ABA accumulation is much higher in nodules than in roots. The effect of ABA on nodule functioning has been shown by exogenous treatment of pea nodules (Gonzalez et al., 2001). ABA treatment decreases the BNF and Lb content declines in parallel with the BNF. However, sucrose synthase activity, another parameter of drought stress effect, is not affected by this treatment suggesting that ABA is not the only player of nodule response to stress. Jasmonic acid (JA), another hormone involved in plant stress response (Reinbothe et al., 2009), has also been shown to be involved in the regulation of nodule functioning (Hause & Schaarschmidt, 2009). Exogenous JA treatment induces an accumulation of lipid peroxides and modifies ascorbate metabolism suggesting that JA could influence nodule senescence (Loscos et al., 2008).

Redox state modifications seem to be a regulatory element of the nodule SIS (Marino et al., 2006). Exogenous treatment with paraquat, which generates ROS, induces the alteration of the GSH and ASC pools toward a more oxidized state. This alteration of the redox state is associated with a diminution of BNF and decrease in Lb content. Moreover, an early decrease in sucrose synthase activity is also detected during the treatment. These results suggest that oxidative stress is involved in the signalling pathway leading to nodule SIS. Interestingly, sucrose synthase seems to be regulated at both the transcriptional and post-translational levels by oxidizing agents such as paraquat (Marino et al., 2008). Finally, genetic modifications allowing decrease and increase of GSH content in the nitrogen fixing zone has shown that BNF and Lb expression level are correlated with GSH content (El Msehli et al., 2011). These results strengthen the idea that cellular redox state plays a crucial role in the regulation of nodule functioning.

Developmental senescence and SIS present different structural and temporal features. At the transcriptomic level, analysis of the expression of 58 genes up regulated during developmental senescence has been performed during dark stress (Perez Guerra et al., 2010). 21 genes are induced during both types of senescence. Amongst these genes, some serine/threonine kinase and some genes involved in metal metabolism (metal transporters and metallothionein) have similar profiles of induction. Nine are up-regulated during both senescence types with different induction levels or transient induction during dark stress. Amongst these genes, cysteine and aspartic proteases are well represented. Finally, 28 genes are up regulated during developmental senescence and not by dark stress. Amongst these genes, proteins associated to proteasome function and vesicular trafficking are not induced during dark stress suggesting partial different regulatory processes between the two types of senescence. In soybean, a screen for genes involved in root nodule senescence has led to the isolation of the senescence-associated nodulin 1 (SAN1) multigene family showing a high homology with plant 2-oxoglutarate-dependent dioxygenases and including two functional genes SAN1A and SAN1B and a pseudogene SAN1C (Webb et al., 2008). Analyses of the steady-state mRNA levels of SAN1A and SAN1B during developmental senescence showed no significant differences for both genes. In contrast, during induced senescence by treatment with nitrate or darkness, SAN1A is down-regulated and SAN1B is up-regulated by both treatments.

Nevertheless, dark stress, drought stress and salt stress may induce specific senescence cascades and transcriptomic analyses will have to be realized to define the similarities and the differences in gene expression patterns in nodules subjected to the different stress.

4. Bacterial mutants and nodule senescence

The microsymbiont, differentiated into bacteroids inside the symbiosome, is not only dependent on "senescent" signals coming from the cytosolic environment of its host plant.

Mutations in some bacteroid genes have an incidence on its lifespan *in planta* and thus on the nodule integrity. However, it is difficult to assess the role of bacteroid genes on nodule senescence due to a lack of genetic tools to investigate this question.

Gene affected	Species	Function	Symbiotic phenotype	Reference
nifH, fixA, fixJ and fixK	*S. meliloti*	Nitrogen fixation	fix⁻, elongated differentiated bacteroids	Maunoury et al., (2010)
lspB	*S. meliloti*	Lipopolysaccharide biosynthesis		
rpoH1	*S. meliloti*	σ²³-like protein , bacterial protection against environmental stresses	fix⁻, reduced number of intracellular bacteria, mixture of normal elongated and abnormal bacteroids	Mitsui, (2004)
gltA	*S. meliloti*	Citrate synthase	fix⁺/⁻ (20%/WT)	Grzemski et al., (2005)
aap bra	*R. leguminosarum*	Branched-chain amino acids transporters	fix⁺/⁻ (30%/WT), fewer smaller bacteroids with a lower DNA content	Prell et al., (2009)
relA	*R. etli*	Stringent response	fix⁺/⁻ (25%/WT), abnormal bacteroids, with an increased size	Moris et al., (2005)
sitA	*S. meliloti*	Manganese uptake	fix⁺/⁻ (25%/WT), mixture of white, slightly pink and pink nodules	Davies et al., (2007)
katA katC	*S. meliloti*	Catalases	fix⁺/⁻ (30%/WT), differentiated bacteroids irregular in shape	Jamet et al., (2003)
gshB	*S. meliloti*	Glutathione synthetase	fix⁺/⁻ (25%/WT), senescence before or after differentiation into bacteroid	Harrison et al., (2005)
gshB	*R. tropici*	Glutathione synthetase	fix⁺/⁻	Muglia et al., (2008)
lsrB1	*S. meliloti*	LysR-type transcriptional regulator	fix⁺/⁻ (30%/WT), mixture of pink and white nodules	Luo et al., (2005)
hfq	*S. meliloti*	RNA chaperone	fix⁺/⁻ (50%/WT), mixture of pink and white nodules	Torres-Quesada et al., (2010)

Table 1. Bacterial genes in which mutations cause early nodule senescence

Until now, there is no global transcriptomic study which could give information on the bacteroid gene expression profile when nitrogen fixing bacteroids turn off to become senescent. Moreover, due to the difficulty in finding a reliable and efficient screen to isolate bacteroid senescent mutants *in planta*, a library of such mutants is still not available. The data described below mostly come from the analysis of the symbiotic phenotype of rhizobial strains affected in one specific gene. The characteristics associated to these analyses include: plant yield, nodule morphology, nitrogen fixation efficiency, and sometimes, ultrastructure of the nodule and of the bacteroid. More recently, some symbiotic phenotype studies also include bacterial genome endoreduplication and transcriptome analyses. This section will mainly focus on the impact of a mutation in the bacterial genome on early nodule senescence and on delayed nodule senescence (Table 1).

4.1 Bacterial mutants and early nodule senescence

The rhizobial mutants that present a symbiotic phenotype are divided into four groups: the nodule deficient mutants impaired in the first steps of infection (nod⁻ fix⁻), the bacterial mutants which induce nodules that present an early nodule senescence phenotype i) blocked in their bacteroid differentiation process (nod⁺ fix⁻), ii) fully differentiated but unable to reduce N_2 (nod⁺ fix⁻) and iii) differentiated into bacteroids less efficient in N_2 fixation compared to the wild-type strain (nod⁺ fix⁺/⁻). Bacterial mutants leading to nodule development abortion due to a defect in bacteroid differentiation such as *bacA* (Saeki, 2011) or *parA* (Liu et al., 2011) mutants will not be presented here. The rhizobial nod⁺ fix⁻ mutants that could differentiate into bacteroids and pass the two transcriptome switch-points encountered during the differentiation of the wild-type bacteroids have been described recently (Maunoury et al., 2010). They are affected in genes encoding for symbiotic function and nitrogen fixation machinery (*nifH*, *fixA*, *fixJ* and *fixK*) or in lipopolysaccharide biosynthesis (*lspB*). The *rpoH1* mutant of *S. meliloti*, impaired in the synthesis of the σ³²-like protein, involved in bacterial protection against environmental stresses has also a nod⁺ fix⁻ symbiotic phenotype in interaction with alfalfa (Mitsui et al., 2004). The bacterial *rpoH1* mutant is still able to elicit nodule formation, efficient plant cell invasion and differentiation into bacteroids. But, the degeneration of bacteroids rapidly occurred in the proximal zone, adjacent to the infection zone, leading to ineffective white nodules associated to an early nodule senescence phenotype. Only the latter group of early nodule senescence mutants (nod⁺ fix⁺/⁻) will be described below. The genes affected in these mutants fall mainly into two categories: genes encoding function involved in carbon and nitrogen metabolism, and genes important for stress adaptation. Genes with other function will also be presented.

4.1.1 Genes encoding function involved in carbon and nitrogen metabolism of the bacteroid and in the nutriment stress response

The host plant supplies bacteroids with metabolites, including dicarboxylic and amino acids, used by the bacteroids to support the reduction of N_2 into ammonia in amounts sufficient for plant growth. To better understand the role of the decarboxylating part of the *S. meliloti* TCA cycle in a nitrogen fixing nodule, Gremski and collaborators (2005) have used an elegant approach (Grzemski et al., 2005). In *S. meliloti*, a mutant in the TCA cycle *gltA* gene encoding citrate synthase forms empty nodules devoid of intracellular bacteria. The *gltA* mutants are clearly unsuitable for experiments to determine whether citrate synthase (CS) is essential during nitrogen fixation in a mature nodule because they have a defect in development that prevents them from forming normal bacteroids. So, the authors constructed temperature-sensitive (ts) mutants in the *S. meliloti* citrate synthase (*gltA*) gene. This allows the formation of nitrogen-fixing nodules at the permissive temperature but, once nodule development was complete, an elevation of root temperature prevents CS xpression. When alfalfa plants infected with the ts mutants were transferred to 30°C, the iodules lost the ability to fix nitrogen. Microscopic examination of the nodules revealed the oss of bacteroids in infected cells and morphological changes that resembled changes seen luring nodule senescence.

These experiments with CS ts mutants showed that CS activity is needed in mature nodules o maintain bacteroid integrity and that removing CS activity via a temperature shift onverts an effective nodule into an empty nodule. This implies that CS is essential for iodule maintenance as well as in the early stages of plant cell invasion.

Another example of early nodule senescence associated to a nutrient defect in *Rhizobium* has)een described recently. Prell and collaborators (2009) have shown that *Rhizobium eguminosarum*, the bacterial partner of peas and broad beans (biovar *viciae*), becomes symbiotic auxotroph for the branched-chain amino acids Leucine, Isoleucine, Valine (LIV) when differentiated into bacteroids in root nodules (Prell et al., 2009). While these bacteria are prototrophs for LIV amino acids as free-living bacteria, they become dependent on the plant as nitrogen-fixing bacteroids, due to a major reduction in gene expression and activity of LIV biosynthetic enzymes. Peas inoculated with bacterial mutants impaired in their apacity to transport LIV, have an early senescent phenotype (nod$^+$, fix$^{+/-}$). Peas are yellow, iave small, pale pink nodules and a dry weight similar to un-inoculated plants. This is correlated with a 70% decrease of the nitrogen fixation capacity for plants inoculated with these mutants compared to the plants inoculated with the *R. leguminosarum* wild-type strain. Thus, plants not only provide a carbon source (dicarboxylic acids) to the bacteroid but also precursors of proteins. The authors have shown that a defect in bacteroid LIV nutrition leads to a reduction in its persistence in plant infected cells, which in turn induces senescence. This means that the plant cell might receive information from the bacteroid in order to sense the fitness of the microsymbiont to maintain or interrupt the symbiotic interaction.

Concerning nutriment stress perceived by the bacteria *in planta*, a mutation in the *relA* gene of *Rhizobium etli*, induces symbiotic defects at the intermediate and/or late stages of the interaction with *Phaseolus vulgaris* (Moris et al., 2005). RelA allows the production of the alarmone (p)ppGpp, which mediates the stringent response in bacteria. This response results in transcriptional down regulation of ribosomal and tRNA genes, upon conditions of amino acid starvation. Despite this role, RelA has been reported to be important for biofilm formation and for interaction of bacteria, pathogenic or beneficial, with their eukaryotic host. Interaction of a *relA* mutant with common bean plants strongly reduces the nitrogen fixation efficiency by 75% and the plant yield. Microscopic studies showed that bacteria differentiated into bacteroids in the symbiosomes were larger in size than the wild-type ones. Thus, in the *R. etli* bacteroids, *relA* plays a role in physiology adaptation and regulation of gene expression. However, the step impaired in the nitrogen fixation defect of a *relA* mutant has not been investigated.

4.1.2 Genes encoding function involved in oxidative stress response

It is generally accepted that symbiotic bacteria are submitted to an oxidative burst released by the host plant during the first steps of infection (Pauly et al., 2006). In addition to this role as a general plant defence mechanism against bacterial invasion, oxidative burst might also play a role in the lifespan of the bacteroid. The high rate of bacteroid respiration necessary to supply energy required for the nitrogen reduction process generates high levels of ROS in the nodule. In legume root nodules, a large amount of H_2O_2 surrounding disintegrating bacteroids in senescent zone IV is detected (Rubio et al, 2004). This reflects the close relationship between oxidative stress and nodule senescence. Indeed, most of the bacterial mutant strains that show a symbiotic nod$^+$ fix$^{+/-}$ phenotype (early senescence) are affected in their antioxidant defence. To escape the stress generated by H_2O_2 and O_2^-, bacteria encode a set of enzymes such as superoxide dismutases, catalases and alkylhydroperoxidases, and also antioxidant molecules such as GSH.

In S. meliloti, disruption of the sitA gene induces a decrease in Mn/Fe SodB activity and a higher sensitivity to ROS (Davies and Walker, 2007). The sitA gene encodes a periplasmic protein involved in manganese uptake. During alfalfa interaction, a sitA mutant is either affected in its infection efficiency leading to small white nodules or is possibly altered in the survival of the differentiated form leading to intermediate nodules, with a slight pink fixing zone, smaller in size than the wild-type nodules. As a consequence, the nitrogenase activity and the plant yield are greatly reduced in these mutant infected plants compared to the plants inoculated with the wild-type bacteria. It is difficult to assess the role of the other superoxide dismutase of S. meliloti (SodA) in the natural senescence process since a sodA mutant failed to differentiate into bacteroids after release into plant cells (Santos et al., 2000).

To cope with H_2O_2 production, S. meliloti possesses three catalases. In free living bacteria, KatA and KatC are encoded by genes mainly transcribed in oxidative stress conditions while the katB gene is constitutively expressed. In 6 week-old nodules of Medicago sativa, KatA is the predominant catalase present in the bacteroids. The katB gene is also expressed in the nitrogen fixation zone III while katC is only transcribed in the infection zone II (Jamet et al., 2003). Single mutant strains of katA, katB or katC genes have no significant impact on nitrogen fixation efficiency of alfalfa nodules containing these mutants compared to those infected with wild-type bacteria. However a katA katC double mutant presents a dramatic decrease of nitrogen fixation capacity (Sigaud et al., 1999), associated with an early senescence of the nodule. These nodules were devoid of a clear zone III, instead the senescent zone IV was adjacent to interzone II-III (Jamet et al., 2003). In most plant cells, bacteria were correctly released from infection threads and were able to differentiate into bacteroids. This shows that efficient detoxification of H_2O_2 by the microsymbiont is essential in the latter steps of bacteroid differentiation leading to nitrogen fixing bacteria.

The antioxidant GSH plays an important role during symbiosis and nodule senescence. This tripeptide is synthesized by a two-step process. In bacteria, GshA catalyses the conjugation of glutamate and cysteine to form γEC and, in a second enzymatic step, GshB completes GSH synthesis by addition of glycine. In S. meliloti, while a gshA mutant is unable to form nodules (nod$^-$ fix$^-$ phenotype), a gshB mutant has a nod$^+$ phenotype coupled to a 75% reduction in the nitrogen fixation capacity (Harrison et al., 2005). In these nodules, bacteria are correctly released from the infection thread into host plant cells and enter into early senescence after differentiation into bacteroids. These data show that, in S. meliloti, GSH is important to maintain bacteroid during symbiotic interaction with alfalfa. This is also true in some determinate-type nodules as the survival of the common bean (P. vulgaris) microsymbiont, Rhizobium tropicii, is dependent on GSH production (Muglia et al., 2008). A gshB mutant has an early senescent pattern associated with increased levels of superoxide accumulation. Expression of this gene in a wild-type background is enhanced at late stage of nodule development, suggesting its antioxidant role against ROS accumulation during nodule senescence. In these species, GSH is important to keep nodules functional over time. In contrast, this does not hold true for *Bradyrhizobium* sp. where disruption of the *gshA* gene does not affect the ability to form effective nodules (Sobrevals et al., 2006). In this latter case, it is possible that the defect in intracellular GSH was compensated for by other compounds acting as antioxidants.

4.1.3 Genes encoding bacterial function involved in the regulation of gene expression

Ninety putative genes encoding LysR-type transcriptional regulators were identified in the *S. meliloti* genome. These regulators are typically 300 amino acids long with an N-terminal DNA binding domain and a C-terminal sensing domain for signal molecules and function as activators or repressors. LysR regulated genes have promoters which contain at least one TN11A motif and are usually divergently transcribed from the LysR regulator (Schell, 1993). To determine the role of LysR regulators in symbiosis, a mutagenesis analysis of all 90 putative *lysR* genes was realized (Luo et al., 2005). This allowed the isolation of the *lsrB1* mutant that presents a symbiotic phenotype. An *lsrB1* mutant was deficient in symbiosis and elicited a mixture of pink (45%) and white (55%) nodules on alfalfa plants. These plants exhibited lower overall nitrogenase activity (30%) than plants inoculated with the wild-type strain. This is consistent with the fact that most of the alfalfa plants inoculated with the *lsrB1* mutant were short (50 to 80% shorter than the plants inoculated with the WT strain) and light green. Cells of the *lsrB1* mutant were recovered from both pink and white nodules, suggesting that *lsrB1* mutants could be blocked either early or late during nodulation. Similar numbers of bacterial cells were recovered from the pink nodules of plants inoculated with the wild-type strain Rm1021 and pink and white nodules from the plants inoculated with the mutants. These findings suggest that the *lsrB1* mutants were able to invade plant cells. The *lsrB* gene is located downstream from the *trxB* gene for thioredoxin reductase, which also participates in the bacterial antioxidant defence. The *trxB* gene is transcribed from its own promoter in the same direction as the *lsrB* gene. The *trxB* promoter contains a nearly perfect recognition site (TN11A) for a LysR regulator so it is possible that LsrB regulates the expression of both the *trxB* and *lsrB* genes. The authors suggest that the early senescence phenotype observed *in planta* could be linked to a defect in detoxification of ROS in *S. meliloti*. However, this has not been demonstrated.

It has been shown recently that the *S. meliloti* RNA chaperone Hfq plays a role in the survival of the microsymbiont within the alfalfa nodule cells (Torres-Quesada et al., 2010). Hfq is considered to act as a global post-transcriptional regulator of gene expression since it interacts with diverse RNA molecules and small non-coding RNAs (sRNA). In free living bacteria, an *hfq* mutant down-regulates 91 genes mostly involved in central carbon metabolism (uptake and utilization of carbon substrates) and up-regulates genes involved in the uptake and catabolism of diverse N compounds. During late interaction with alfalfa (30 days post-infection), plants inoculated with the *hfq* mutant strain are composed of 60% of white non fixing nodules and 40% of pink fixing nodules. Thus, the plant yield was 64% of that of the wild-type-inoculated plants. Histological analysis of the white nodules revealed that the bacteroid differentiation was efficient but the bacteroid-infected tissues were restricted to the interzone II-III since the zone III was replaced by a large senescent zone IV. Indeed, an Hfq impaired mutant showed a premature senescent phenotype. The authors proposed that this phenotype could be linked to a defect in intracellular survival under prolonged stress present in the plant cell environment.

4.2 Bacterial mutants and delayed senescence

The delayed senescent bacterial mutants might have a nod[+] phenotype associated with a fix[+] phenotype for a period longer than the natural fixing period associated with the wild type bacteria in interaction with its host plant. Thus, such mutants are obligate differentiated bacteroids.

Compared to the data connected with the consequences of bacterial gene inactivation on early nodule senescence, little information on the role of bacterial mutants on delayed nodule senescence are available. Knowing that most of the mutations that induce an acceleration of senescence affect genes involved in ROS detoxification, in bacterial fitness, in import and/or processing of carbon skeletons, amino acids and antioxidants, we might suspect that a delayed senescence bacterial mutant should have a gain of function rather than an invalidated one. In that sense, it is possible that an increase in synthesis and/or activity of molecules involved in stress resistance, especially to ROS, should improve the bacteroid lifetime in the symbiosome and thus should enhance the nitrogen fixing period. This aspect of the role of bacterial genes in the functional life of symbiotic fixing nodules remains to be explored. However, one encouraging study sustains this postulate (Redondo et al., 2009). In fact, the authors of this work have overexpressed a cyanobacteria *Anabaena variabilis* gene encoding flavodoxin in *S. meliloti*. Knowing that natural senescence-inducing signals from the plant leads to a decrease in antioxidant content and thus an increase in ROS accumulation in an irreversible manner, they analyse the consequences of the over-expression of this flavodoxin protein involved in the response to oxidative stress. They have shown that the decline of nitrogenase activity was delayed and that the structural and ultrastructural modifications associated with nodule senescence had a later onset in flavodoxin-expressing nodules. Lipid peroxidation, a marker of senescence, was significantly reduced and the oxidative balance was improved in comparison to the control nodules. In conclusion, flavodoxin over-expression had an impact on bacteroid antioxidant metabolism, leading to delayed senescence.

In conclusion, we can propose that bacteroids inside the nodule infected cells are not only tributary from the plant to initiate nodule senescence. Genes encoding proteins implicated in bacterial nutrition and stress response are also essential since mutations in these genes alter the fitness of the differentiated bacteroids. In turn, this leads to the death of the microsymbiont followed by the senescence of the plant cells and nodule. Future aspects on the role of bacteroid genes on senescence should include the development of bacteroid genetic tools. The over-expression of pertinent genes specifically in the bacteroid or the conditional invalidation of rhizobial genes after bacteroid differentiation will be important to define senescence-related genes.

5. Perspectives

The data summarized in this review indicate that one of the general physiological features of nodule senescence is the decrease in nitrogen fixation efficiency. This diminution may be related to plant and/or bacterial-dependent factors. However whereas this diminution of the nitrogen fixation efficiency is observed during both developmental and stress induced-nodule senescence, the progression of senescence symptoms seems to be differentially controlled. Indeed, even if common general features have been described during nodule senescence, the few available microscopic and transcriptomic analyses show that nodule senescence may occur differently in developmental senescence and SIS. Moreover, the senescence occurring under different environmental stress conditions such as dark stress, drought stress or nitrate treatment may also involve different genetic and physiological programs. In this context, more detailed spatiotemporal analysis of the multiple senescence processes will have to be performed to determine the similarities and the differences between the various senescence processes. Microarray analysis or "Whole Transcriptome Shotgun Sequencing" will be valuable tools to analyse the transcriptome modifications occurring under the different senescence processes (Lister et al., 2009). In conditions in which the senescence process does not appear to be a homogenous process such as in developmental senescence, laser capture microdissection (Barcala et al., 2009) will allow the analysis of transcriptomic patterns of senescent infected cells. The development of two legume model systems, M. truncatula (http://www.medicago.org/) and L. japonicus (http://www.lotusjaponicus.org/) will facilitate an efficient analysis of the senescence process by developing tools dedicated to cell biology, genetic and transcriptomic analyses.

The role of the bacterial partner needs also to be clarified. Indeed, whereas senescence phenotypes are observed in nodules obtained with bacterial mutants affected in their nitrogen fixation efficiency, the regulation of this bacterial-induced senescence has not yet been studied. The abortion of nodule development when using such bacterial mutants suggests that the interaction may switch from a compatible to an incompatible interaction. In this context, the molecular events which trigger this switch still need to be defined. Nevertheless, the work with the flavodoxin overexpressing-bacterial strain showed that the symbiotic interaction may also be improved to resist to the various endogenous and environmental stress conditions. The construction of plant and bacteria with higher resistance to environmental stress (Zurbriggen et al., 2008) may be an interesting opportunity to increase the benefit from an efficient BNF in agronomy.

6. References

Alesandrini, F., Mathis, R., Van de Sype, G., Herouart, D. & Puppo, A. (2003). Possible roles for a cysteine protease and hydrogen peroxide in soybean nodule development and senescence. New Phytol, Vol. 158, No. 1, pp. 131-138

Apel, K. & Hirt, H. (2004). Reactive oxygen species: metabolism, oxidative stress, and signal transduction. Annu Rev Plant Biol, Vol. 55, pp. 373-399

Appleby, C. A. (1984). Leghemoglobin and Rhizobium respiration. Annu Rev Plant Physiol Plant Mol Biol., Vol. 35, pp. 443-478

Aydi, S., Drevon, J. J. & Abdelly, C. (2004). Effect of salinity on root-nodule conductance to the oxygen diffusion in the Medicago truncatula-Sinorhizobium meliloti symbiosis. Plant Physiol Biochem, Vol. 42, No. 10, pp. 833-840

Barcala, M., Garcia, A., Cabrera, J., Casson, S., Lindsey, K., Favery, B., Garcia-Casado, G., Solano, R., Fenoll, C. & Escobar, C. (2009). Early transcriptomic events in microdissected Arabidopsis nematode-induced giant cells. Plant J, Vol. 61, No. 4, pp. 698-712

Becana, M. & Klucas, R. V. (1992). Transition metals in legume root nodules: iron-dependent free radical production increases during nodule senescence. *Proc Natl Acad Sci U S A*, Vol. 89, No. 19, pp. 8958-8962

Becana, M., Matamoros, M. A., Udvardi, M. & Dalton, D. A. (2010). Recent insights into antioxidant defenses of legume root nodules. *New Phytol*, Vol. 188, No. 4, pp. 960-976

Becker, A., Berges, H., Krol, E., Bruand, C., Ruberg, S., Capela, D., Lauber, E., Meilhoc, E., Ampe, F., de Bruijn, F. J., Fourment, J., Francez-Charlot, A., Kahn, D., Kuster, H., Liebe, C., Puhler, A., Weidner, S. & Batut, J. (2004). Global changes in gene expression in Sinorhizobium meliloti 1021 under microoxic and symbiotic conditions. *Mol Plant Microbe Interact*, Vol. 17, No. 3, pp. 292-303

Bekki, A., Trinchant, J.-C. & Rigaud, J. (1987). Nitrogen fixation (C_2H_2 reduction) by Medicago nodules and bacteroids under sodium chloride stress. *Physiologia Plantarum*, Vol. 71, No. 1, pp. 61-67

Ben Salah, I., Albacete, A., Martinez Andujar, C., Haouala, R., Labidi, N., Zribi, F., Martinez, V., Perez-Alfocea, F. & Abdelly, C. (2009). Response of nitrogen fixation in relation to nodule carbohydrate metabolism in Medicago ciliaris lines subjected to salt stress. *J Plant Physiol*, Vol. 166, No. 5, pp. 477-488

Benedito, V. A., Torres-Jerez, I., Murray, J. D., Andriankaja, A., Allen, S., Kakar, K., Wandrey, M., Verdier, J., Zuber, H., Ott, T., Moreau, S., Niebel, A., Frickey, T., Weiller, G., He, J., Dai, X., Zhao, P. X., Tang, Y. & Udvardi, M. K. (2008). A gene expression atlas of the model legume Medicago truncatula. *Plant J*, Vol. 55, No. 3, pp. 504-513

Bethlenfalvay, G. J. & Phillips, D. A. (1977). Ontogenetic Interactions between Photosynthesis and Symbiotic Nitrogen Fixation in Legumes. *Plant Physiol*, Vol. 60, No. 3, pp. 419-421

Bobik, C., Meilhoc, E. & Batut, J. (2006). FixJ: a major regulator of the oxygen limitation response and late symbiotic functions of Sinorhizobium meliloti. *J Bacteriol*, Vol. 188, No. 13, pp. 4890-4902

Broughton, W. J., Jabbouri, S. & Perret, X. (2000). Keys to symbiotic harmony. *J Bacteriol* Vol. 182, No. 20, pp. 5641-5652

Capela, D., Filipe, C., Bobik, C., Batut, J. & Bruand, C. (2006). Sinorhizobium meliloti differentiation during symbiosis with alfalfa: a transcriptomic dissection. *Mol Plant Microbe Interact*, Vol. 19, No. 4, pp. 363-372

Chang, C., Damiani, I., Puppo, A. & Frendo, P. (2009). Redox changes during the legume-rhizobium symbiosis. *Mol Plant*, Vol. 2, No. 3, pp. 370-377

Clement, M., Lambert, A., Herouart, D. & Boncompagni, E. (2008). Identification of new up-regulated genes under drought stress in soybean nodules. *Gene*, Vol. 426, No. 1, pp. 15-22

Coba de la Pena, T., Carcamo, C. B., Almonacid, L., Zaballos, A., Lucas, M. M., Balomenos, D. & Pueyo, J. J. (2008). A cytokinin receptor homologue is induced during root nodule organogenesis and senescence in Lupinus albus L. *Plant Physiol Biochem*, Vol. 46, No. 2, pp. 219-225

Crespi, M. & Frugier, F. (2008). De novo organ formation from differentiated cells: root nodule organogenesis. *Sci Signal*, Vol. 1, No. 49, pp. re11

Cutler, S. R., Rodriguez, P. L., Finkelstein, R. R. & Abrams, S. R. (2010). Abscisic acid: emergence of a core signaling network. *Annu Rev Plant Biol,* Vol. 61, pp. 651-679

Davies, B. W. & Walker, G. C. (2007). Disruption of sitA compromises Sinorhizobium meliloti for manganese uptake required for protection against oxidative stress. *J Bacteriol,* Vol. 189, No. 5, pp. 2101-2109

De Michele, R., Formentin, E., Todesco, M., Toppo, S., Carimi, F., Zottini, M., Barizza, E., Ferrarini, A., Delledonne, M., Fontana, P. & Lo Schiavo, F. (2009). Transcriptome analysis of Medicago truncatula leaf senescence: similarities and differences in metabolic and transcriptional regulations as compared with Arabidopsis, nodule senescence and nitric oxide signalling. *New Phytol,* Vol. 181, No. 3, pp. 563-575

Del Castillo, L. D., Hunt, S. & Layzell, D. B. (1994). The role of oxygen in the regulation of nitrogenase activity in drought-stressed soybean nodules. *Plant Physiol,* Vol. 106, No. 3, pp. 949-955

Downie, J. A. (2005). Legume haemoglobins: symbiotic nitrogen fixation needs bloody nodules. *Curr Biol,* Vol. 15, No. 6, pp. R196-198

Duc, G. & Messager, A. (1989). Mutagenesis of pea (Pisum sativum L.) and the isolation of mutants for nodulation and nitrogen fixation. *Plant Science,* Vol. 60, No. 2, pp. 207-213

El Msehli, S., Lambert, A., Baldacci-Cresp, F., Hopkins, J., Boncompagni, E., Smiti, S. A., Herouart, D. & Frendo, P. (2011). Crucial role of (homo)glutathione in nitrogen fixation in Medicago truncatula nodules. *New Phytol,* in press

El Yahyaoui, F., Kuster, H., Ben Amor, B., Hohnjec, N., Puhler, A., Becker, A., Gouzy, J., Vernie, T., Gough, C., Niebel, A., Godiard, L. & Gamas, P. (2004). Expression profiling in Medicago truncatula identifies more than 750 genes differentially expressed during nodulation, including many potential regulators of the symbiotic program. *Plant Physiol,* Vol. 136, No. 2, pp. 3159-3176

Escuredo, P. R., Minchin, F. R., Gogorcena, Y., Iturbe-Ormaetxe, I., Klucas, R. V. & Becana, M. (1996). Involvement of Activated Oxygen in Nitrate-Induced Senescence of Pea Root Nodules. *Plant Physiol,* Vol. 110, No. 4, pp. 1187-1195

Evans, P. J., Gallesi, D., Mathieu, C., Hernandez, M. J., de Felipe, M., Halliwell, B. & Puppo, A. (1999). Oxidative stress occurs during soybean nodule senescence. *Planta,* Vol. 208, No.1, pp. 73-79

Fedorova, M., van de Mortel, J., Matsumoto, P. A., Cho, J., Town, C. D., VandenBosch, K. A., Gantt, J. S. & Vance, C. P. (2002). Genome-wide identification of nodule-specific transcripts in the model legume Medicago truncatula. *Plant Physiol,* Vol. 130, No. 2, pp. 519-537

Franssen, H. J., Vijn, I., Yang, W. C. & Bisseling, T. (1992). Developmental aspects of the Rhizobium-legume symbiosis. *Plant Mol Biol,* Vol. 19, No. 1, pp. 89-107

Fridovich, I. (1986). Biological effects of the superoxide radical. *Arch Biochem Biophys,* Vol. 247, No. 1, pp. 1-11

Fujikake, H. - Yamazaki, A., Ohtake, N., Sueyoshi, K., Matsuhashi, S., Ito, T., Mizuniwa, C., Kume, T., Hashimoto, S., Ishioka, N.-S., Watanabe, S., Osa, A., Sekine, T., Uchida, H., Tsuji, A. & Ohyama, T. (2003). Quick and reversible inhibition of soybean root nodule growth by nitrate involves a decrease in sucrose supply to nodules. *J Exp Bot,* Vol. 54, No. 386, pp. 1379-1388

Galibert, F., Finan, T. M., Long, S. R., Puhler, A., Abola, P., Ampe, F., Barloy-Hubler, F., Barnett, M. J., Becker, A., Boistard, P., Bothe, G., Boutry, M., Bowser, L., Buhrmester, J., Cadieu, E., Capela, D., Chain, P., Cowie, A., Davis, R. W., Dreano, S., Federspiel, N. A., Fisher, R. F., Gloux, S., Godrie, T., Goffeau, A., Golding, B., Gouzy, J., Gurjal, M., Hernandez-Lucas, I., Hong, A., Huizar, L., Hyman, R. W., Jones, T., Kahn, D., Kahn, M. L., Kalman, S., Keating, D. H., Kiss, E., Komp, C., Lelaure, V., Masuy, D., Palm, C., Peck, M. C., Pohl, T. M., Portetelle, D., Purnelle B., Ramsperger, U., Surzycki, R., Thebault, P., Vandenbol, M., Vorholter, F. J., Weidner, S., Wells, D. H., Wong, K., Yeh, K. C. & Batut, J. (2001). The composite genome of the legume symbiont Sinorhizobium meliloti. *Science*, Vol. 293, No. 5530 pp. 668-672

Galvez, L., Gonzalez, E. M. & Arrese-Igor, C. (2005). Evidence for carbon flux shortage and strong carbon/nitrogen interactions in pea nodules at early stages of water stress. *Exp Bot*, Vol. 56, No. 419, pp. 2551-2561

Gepstein, S. (2004). Leaf senescence--not just a 'wear and tear' phenomenon. *Genome Biol*, Vol. 5, No. 3, pp. 212

Gogorcena, Y., Iturbe-Ormaetxe, I., Escuredo, P. R. & Becana, M. (1995). Antioxidant defenses against activated oxygen in pea nodules subjected to water stress. *Plant Physiol*, Vol. 108, No. 2, pp. 753-759

Gogorcena, Y., Gordon, A. J., Escuredo, P. R., Minchin, F. R., Witty, J. F., Moran, J. F. & Becana, M. (1997). N2 fixation, carbon metabolism and oxidative damage in nodules of dark stressed Common Bean plants. *Plant Physiol*, Vol. 113, No.4, pp. 1193-1201

Gonzalez, E. M., Gordon, A. J., James, C. L. & Arrese-Igor, C. (1995). The role of sucrose synthase in the response of soybean nodules to drought. *J Ex Bot*, Vol. 46, No. 10 pp. 1515-1523

Gonzalez, E. M., Galvez, L. & Arrese-Igor, C. (2001). Abscisic acid induces a decline in nitrogen fixation that involves leghaemoglobin, but is independent of sucrose synthase activity. *J Exp Bot*, Vol. 52, No. 355, pp. 285-293

Gordon, A. J., Minchin, F. R., Skot, L. & James, C. L. (1997). Stress-induced declines in soybean N2 fixation are related to nodule sucrose synthase activity. *Plant Physiol*, Vol. 114, No. 3, pp. 937-946

Gordon, A. J., Skot, L., James, C. L. & Minchin, F. R. (2002). Short-term metabolic responses of soybean root nodules to nitrate. *J Exp Bot*, Vol. 53, No. 368, pp. 423-428

Groten, K., Vanacker, H., Dutilleul, C., Bastian, F., Bernard, S., Carzaniga, R. & Foyer, C. H. (2005). The roles of redox processes in pea nodule development and senescence. *Plant, Cell & Environment*, Vol. 28, No. 10, pp. 1293-1304

Groten, K., Dutilleul, C., van Heerden, P. D., Vanacker, H., Bernard, S., Finkemeier, I., Dietz, K. J. & Foyer, C. H. (2006). Redox regulation of peroxiredoxin and proteinases by ascorbate and thiols during pea root nodule senescence. *FEBS Lett*, Vol. 580, No. 5, pp. 1269-1276

Grzemski, W., Akowski, J. P. & Kahn, M. L. (2005). Probing the Sinorhizobium meliloti-alfalfa symbiosis using temperature-sensitive and impaired-function citrate synthase mutants. *Mol Plant Microbe Interact*, Vol. 18, No. 2, pp. 134-141

Guerin, V., Trinchant, J. C. & Rigaud, J. (1990). Nitrogen Fixation (C(2)H(2) Reduction) by Broad Bean (Vicia faba L.) Nodules and Bacteroids under Water-Restricted Conditions. *Plant Physiol*, Vol. 92, No. 3, pp. 595-601

Gunther, C., Schlereth, A., Udvardi, M. & Ott, T. (2007). Metabolism of reactive oxygen species is attenuated in leghemoglobin-deficient nodules of Lotus japonicus. *Mol Plant Microbe Interact*, Vol. 20, No. 12, pp. 1596-1603

Harrison, J., Jamet, A., Muglia, C. I., Van de Sype, G., Aguilar, O. M., Puppo, A. & Frendo, P. (2005). Glutathione plays a fundamental role in growth and symbiotic capacity of Sinorhizobium meliloti. *J Bacteriol*, Vol. 187, No. 1, pp. 168-174

Hause, B. & Schaarschmidt, S. (2009). The role of jasmonates in mutualistic symbioses between plants and soil-born microorganisms. *Phytochemistry*, Vol. 70, No. 13-14, pp. 1589-1599

He, J., Benedito, V. A., Wang, M., Murray, J. D., Zhao, P. X., Tang, Y. & Udvardi, M. K. (2009). The Medicago truncatula gene expression atlas web server. *BMC Bioinformatics*, Vol. 10, pp. 441

Horchani, F., Prevot, M., Boscari, A., Evangelisti, E., Meilhoc, E., Bruand, C., Raymond, P., Boncompagni, E., Aschi-Smiti, S., Puppo, A. & Brouquisse, R. (2011). Both plant and bacterial nitrate reductases contribute to nitric oxide production in Medicago truncatula nitrogen-fixing nodules. *Plant Physiol*, Vol. 155, No. 2, pp. 1023-1036

Ivanov, S., Fedorova, E. & Bisseling, T. (2010). Intracellular plant microbe associations: secretory pathways and the formation of perimicrobial compartments. *Curr Opin Plant Biol*, Vol. 13, No. 4, pp. 372-377

Jamet, A., Sigaud, S., Van de Sype, G., Puppo, A. & Herouart, D. (2003). Expression of the bacterial catalase genes during Sinorhizobium meliloti-Medicago sativa symbiosis and their crucial role during the infection process. *Mol Plant Microbe Interact*, Vol. 16, No. 3, pp. 217-225

Jebara, S., Jebara, M., Limam, F. & Aouani, M. E. (2005). Changes in ascorbate peroxidase, catalase, guaiacol peroxidase and superoxide dismutase activities in common bean (*Phaseolus vulgaris*) nodules under salt stress. *J Plant Physiol*, Vol. 162, No. 8, pp. 929-936

Jeudy, C., Ruffel, S., Freixes, S., Tillard, P., Santoni, A. L., Morel, S., Journet, E. P., Duc, G., Gojon, A., Lepetit, M. & Salon, C. (2009). Adaptation of Medicago truncatula to nitrogen limitation is modulated via local and systemic nodule developmental responses. *New Phytol*, Vol. 185, No. 3, pp. 817-828

Jones, K. M., Kobayashi, H., Davies, B. W., Taga, M. E. & Walker, G. C. (2007). How rhizobial symbionts invade plants: the Sinorhizobium-Medicago model. *Nat Rev Microbiol*, Vol. 5, No. 8, pp. 619-633

Kaneko, T., Nakamura, Y., Sato, S., Asamizu, E., Kato, T., Sasamoto, S., Watanabe, A., Idesawa, K., Ishikawa, A., Kawashima, K., Kimura, T., Kishida, Y., Kiyokawa, C., Kohara, M., Matsumoto, M., Matsuno, A., Mochizuki, Y., Nakayama, S., Nakazaki, N., Shimpo, S., Sugimoto, M., Takeuchi, C., Yamada, M. & Tabata, S. (2000). Complete genome structure of the nitrogen-fixing symbiotic bacterium Mesorhizobium loti. *DNA Res*, Vol. 7, No. 6, pp. 331-338

Kaneko, T., Nakamura, Y., Sato, S., Minamisawa, K., Uchiumi, T., Sasamoto, S., Watanabe, A., Idesawa, K., Iriguchi, M., Kawashima, K., Kohara, M., Matsumoto, M., Shimpo, S., Tsuruoka, H., Wada, T., Yamada, M. & Tabata, S. (2002). Complete genomic sequence of nitrogen-fixing symbiotic bacterium Bradyrhizobium japonicum USDA110. *DNA Res*, Vol. 9, No. 6, pp. 189-197

Kardailsky, I. V. & Brewin, N. J. (1996). Expression of cysteine protease genes in pea nodule development and senescence. *Mol Plant Microbe Interact*, Vol. 9, No. 8, pp. 689-695

Karunakaran, R., Ramachandran, V. K., Seaman, J. C., East, A. K., Mouhsine, B., Mauchline, T. H., Prell, J., Skeffington, A. & Poole, P. S. (2009). Transcriptomic analysis of Rhizobium leguminosarum biovar viciae in symbiosis with host plants Pisum sativum and Vicia cracca. *J Bacteriol*, Vol. 191, No. 12, pp. 4002-4014

Kereszt, A., Mergaert, P., Maroti, G. & Kondorosi, E. (2011). Innate immunity effectors and virulence factors in symbiosis. *Current opinion in microbiology*, Vol. 14, No. 1, pp. 76-81

Kondorosi, E., Roudier, F. & Gendreau, E. (2000). Plant cell-size control: growing by ploidy? *Curr Opin Plant Biol*, Vol. 3, No. 6, pp. 488-492

Krusell, L., Madsen, L. H., Sato, S., Aubert, G., Genua, A., Szczyglowski, K., Duc, G., Kaneko, T., Tabata, S., de Bruijn, F., Pajuelo, E., Sandal, N. & Stougaard, J. (2002). Shoot control of root development and nodulation is mediated by a receptor-like kinase. *Nature*, Vol. 420, No. 6914, pp. 422-426

Kuster, H., Vieweg, M. F., Manthey, K., Baier, M. C., Hohnjec, N. & Perlick, A. M. (2007). Identification and expression regulation of symbiotically activated legume genes. *Phytochemistry*, Vol. 68, No. 1, pp. 8-18

Kuzma, M. M., Topunov, A. F. & Layzell, D. B. (1995). Effects of Temperature on Infected Cell O2 Concentration and Adenylate Levels in Attached Soybean Nodules. *Plant Physiol*, Vol. 107, No. 4, pp. 1209-1216

Larrainzar, E., Wienkoop, S., Weckwerth, W., Ladrera, R., Arrese-Igor, C. & Gonzalez, E. M. (2007). *Medicago truncatula* root nodule proteome analysis reveals differential plant and bacteroid responses to drought stress. *Plant Physiol*, Vol. 144, No. 3, pp. 1495-1507

Lawn, R. J. & Brun, W. A. (1977). Symbiotic Nitrogen Fixation in Soybeans. I. Effect of Photosynthetic Source-Sink Manipulations1. *Crop Sci.*, Vol. 14, No. 1, pp. 11

Lehtovaara, P. & Perttila, U. (1978). Bile-pigment formation from different leghaemoglobins. Methine-bridge specificity of coupled oxidation. *Biochem J*, Vol. 176, No. 2, pp. 359-364

Li, Y., Zhou, L., Li, Y., Chen, D., Tan, X., Lei, L. & Zhou, J. (2008). A nodule-specific plant cysteine proteinase, AsNODF32, is involved in nodule senescence and nitrogen fixation activity of the green manure legume Astragalus sinicus. *New Phytol*, Vol. 180, No. 1, pp. 185-192

Limpens, E., Ivanov, S., van Esse, W., Voets, G., Fedorova, E. & Bisseling, T. (2009). Medicago N2-fixing symbiosomes acquire the endocytic identity marker Rab7 but delay the acquisition of vacuolar identity. *Plant Cell*, Vol. 21, No. 9, pp. 2811-2828

Lister, R., Gregory, B. D. & Ecker, J. R. (2009). Next is now: new technologies for sequencing of genomes, transcriptomes, and beyond. *Curr Opin Plant Biol*, Vol. 12, No. 2, pp. 107-118

Liu, C. T., Lee, K. B., Wang, Y. S., Peng, M. H., Lee, K. T., Suzuki, S., Suzuki, T. & Oyaizu, H. (2011). Involvement of the azorhizobial chromosome partition gene (parA) in the onset of bacteroid differentiation during Sesbania rostrata stem nodule development. *Applied and environmental microbiology*, Vol. 77, No. 13, pp. 4371-4382

Lodwig, E. & Poole, P. S. (2003). Metabolism of Rhizobium bacteroids. *Crit Rev Plant Sci*, Vol. 22, No.1, pp. 37-78

Lohman, K. N., Gan, S., John, M. C. & Amasino, R. M. (1994). Molecular analysis of natural leaf senescence in Arabidopsis thaliana. *Physiologia Plantarum*, Vol. 92, No. 2, pp. 322

Long, S. R. (2001). Genes and signals in the rhizobium-legume symbiosis. *Plant Physiol*, Vol. 125, No. 1, pp. 69-72

Lopez, M., Herrera-Cervera, J. A., Iribarne, C., Tejera, N. A. & Lluch, C. (2008). Growth and nitrogen fixation in *Lotus japonicus* and *Medicago truncatula* under NaCl stress: Nodule carbon metabolism. *J Plant Physiol*, Vol. 165, No. 6, pp. 641-650

Lopez, M., Tejera, N. A. & Lluch, C. (2009). Validamycin A improves the response of Medicago truncatula plants to salt stress by inducing trehalose accumulation in the root nodules. *J Plant Physiol*, Vol. 166, No. 11, pp. 1218-1222

Loscos, J., Matamoros, M. A. & Becana, M. (2008). Ascorbate and homoglutathione metabolism in common bean nodules under stress conditions and during natural senescence. *Plant Physiol*, Vol. 146, No. 3, pp. 1282-1292

Lucas, M. M., Van de Sype, G., Hérouart, D., Hernandez, M. J., Puppo, A. & de Felipe, M. R. (1998). Immunolocalization of ferritin in determinate and indeterminate legume root nodules. *Protoplasma*, Vol. 204, No. 1, pp. 61-70

Luo, L., Yao, S. Y., Becker, A., Ruberg, S., Yu, G. Q., Zhu, J. B. & Cheng, H. P. (2005). Two new Sinorhizobium meliloti LysR-type transcriptional regulators required for nodulation. *J Bacteriol*, Vol. 187, No. 13, pp. 4562-4572

L'Taief, B., Sifi, B., Zaman-Allah, M., Drevon, J. J. & Lachaal, M. (2007). Effect of salinity on root-nodule conductance to the oxygen diffusion in the *Cicer arietinum-Mesorhizobium ciceri* symbiosis. *J Plant Physiol*, Vol. 164, No. 8, pp. 1028-1036

Malik, N. S., Pfeiffer, N. E., Williams, D. R. & Wagner, F. W. (1981). Peptidohydrolases of Soybean Root Nodules: Identification, separation and partial characterisation of enzymes from bacteroid-free extracts. *Plant Physiol*, Vol. 68, No. 2, pp. 386-392

Marino, D., Gonzalez, E. M. & Arrese-Igor, C. (2006). Drought effects on carbon and nitrogen metabolism of pea nodules can be mimicked by paraquat: evidence for the occurrence of two regulation pathways under oxidative stresses. *J Exp Bot*, Vol. 57, No. 3, pp. 665-673

Marino, D., Frendo, P., Ladrera, R., Zabalza, A., Puppo, A., Arrese-Igor, C. & Gonzalez, E. M. (2007). Nitrogen fixation control under drought stress. Localized or systemic? *Plant Physiol*, Vol. 143, No. 4, pp. 1968-1974

Marino, D., Hohnjec, N., Kuster, H., Moran, J. F., Gonzalez, E. M. & Arrese-Igor, C. (2008). Evidence for transcriptional and post-translational regulation of sucrose synthase in pea nodules by the cellular redox state. *Mol Plant Microbe Interact*, Vol. 21, No. 5, pp. 622-630

Marino, D., Pucciariello, C., Puppo, A. & Frendo, P. (2009).The Redox State, a Referee of the Legume-Rhizobia Symbiotic Game. In *Advances in Botanical Research: Oxidative Stress and Redox Regulation in Plants*, J. P. Jacquot, pp. 115-151, Burlington: Academic Press

Matamoros, M. A., Baird, L. M., Escuredo, P. R., Dalton, D. A., Minchin, F. R., Iturbe-Ormaetxe, I., Rubio, M. C., Moran, J. F., Gordon, A. J. & Becana, M. (1999). Stress induced legume root nodule senescence. Physiological, biochemical and structural alterations. *Plant Physiol*, Vol. 121, No.1, pp. 97-112

Maunoury, N., Kondorosi, A., Kondorosi, E. & Mergaert, P. (2008).Cell biology of nodule infection and development. In *Nitrogen-fixing Leguminous Symbioses* E. K. James, J. I. Sprent, W. E. Dilworth and N. W.E., pp. 153–189, Springer, the Netherlands

Maunoury, N., Redondo-Nieto, M., Bourcy, M., Van de Velde, W., Alunni, B., Laporte, P., Durand, P., Agier, N., Marisa, L., Vaubert, D., Delacroix, H., Duc, G., Ratet, P., Aggerbeck, L., Kondorosi, E. & Mergaert, P. (2010). Differentiation of symbiotic cells and endosymbionts in Medicago truncatula nodulation are coupled to two transcriptome-switches. *PLoS One*, Vol. 5, No. 3, pp. e9519

Mergaert, P., Nikovics, K., Kelemen, Z., Maunoury, N., Vaubert, D., Kondorosi, A. & Kondorosi, E. (2003). A novel family in Medicago truncatula consisting of more than 300 nodule-specific genes coding for small, secreted polypeptides with conserved cysteine motifs. *Plant Physiol*, Vol. 132, No. 1, pp. 161-173

Mergaert, P., Uchiumi, T., Alunni, B., Evanno, G., Cheron, A., Catrice, O., Mausset, A. E., Barloy-Hubler, F., Galibert, F., Kondorosi, A. & Kondorosi, E. (2006). Eukaryotic control on bacterial cell cycle and differentiation in the Rhizobium-legume symbiosis. *Proc Natl Acad Sci U S A*, Vol. 103, No. 13, pp. 5230-5235

Mhadhbi, H., Fotopoulos, V., Mylona, P. V., Jebara, M., Elarbi Aouani, M. & Polidoros, A. N. (2011). Antioxidant gene-enzyme responses in Medicago truncatula genotypes with different degree of sensitivity to salinity. *Physiol Plant*, Vol. 141, No. 3, pp. 201-214

Minchin, F. R., Minguez, M. I., Sheehy, J. E., Witty, J. F. & SkÃ~T, L. (1986). Relationships Between Nitrate and Oxygen Supply in Symbiotic Nitrogen Fixation by White Clover. *J Ex Bot*, Vol. 37, No. 8, pp. 1103-1113

Minchin, F. R., Becana, M. & Sprent, J. I. (1989). Short-term inhibition of legume N_2 fixation by nitrate. II. Nitrate effects on nodule oxygen diffusion. *Planta*, Vol. 180, No.1, pp. 46-52

Mitsui, H., Sato, T., Sato, Y., Ito, N. & Minamisawa, K. (2004). Sinorhizobium meliloti RpoH1 is required for effective nitrogen-fixing symbiosis with alfalfa. *Mol Genet Genomics*, Vol. 271, No. 4, pp. 416-425

Moris, M., Braeken, K., Schoeters, E., Verreth, C., Beullens, S., Vanderleyden, J. & Michiels, J. (2005). Effective symbiosis between Rhizobium etli and Phaseolus vulgaris requires the alarmone ppGpp. *J Bacteriol*, Vol. 187, No. 15, pp. 5460-5469

Mortier, V., Holsters, M. & Goormachtig, S. (2011). Never too many? How legumes control nodule numbers. *Plant Cell Environ*, in press

Muglia, C., Comai, G., Spegazzini, E., Riccillo, P. M. & Aguilar, O. M. (2008). Glutathione produced by Rhizobium tropici is important to prevent early senescence in common bean nodules. *FEMS microbiology letters*, Vol. 286, No. 2, pp. 191-198

Naito, Y., Fujie, M., Usami, S., Murooka, Y. & Yamada, T. (2000). The involvement of a cysteine proteinase in the nodule development in Chinese milk vetch infected with Mesorhizobium huakuii subsp. rengei. *Plant Physiol*, Vol. 124, No. 3, pp. 1087-1096

Naya, L., Ladrera, R., Ramos, J., Gonzalez, E. M., Arrese-Igor, C., Minchin, F. R. & Becana, M. (2007). The response of carbon metabolism and antioxidant defenses of alfalfa nodules to drought stress and to the subsequent recovery of plants. *Plant Physiol*, Vol. 144, No. 2, pp. 1104-1114

Novak, K. (2010). Early action of pea symbiotic gene NOD3 is confirmed by adventitious root phenotype. *Plant Sci*, Vol. 179, No. 5, pp. 472-478

)ldroyd, G. E., Murray, J. D., Poole, P. S. & Downie, J. A. (2011). The Rules of Engagement in the Legume-Rhizobial Symbiosis. *Annu Rev Genet*, Vol.45, in press

)tt, T., van Dongen, J. T., Gunther, C., Krusell, L., Desbrosses, G., Vigeolas, H., Bock, V., Czechowski, T., Geigenberger, P. & Udvardi, M. K. (2005). Symbiotic leghemoglobins are crucial for nitrogen fixation in legume root nodules but not for general plant growth and development. *Curr Biol*,Vol. 15, No. 6, pp. 531-535

)tt, T., Sullivan, J., James, E. K., Flemetakis, E., Gunther, C., Gibon, Y., Ronson, C. & Udvardi, M. (2009). Absence of symbiotic leghemoglobins alters bacteroid and plant cell differentiation during development of Lotus japonicus root nodules. *Mol Plant Microbe Interact*, Vol. 22, No. 7, pp. 800-808

'auly, N., Pucciariello, C., Mandon, K., Innocenti, G., Jamet, A., Baudouin, E., Herouart, D., Frendo, P. & Puppo, A. (2006). Reactive oxygen and nitrogen species and glutathione: key players in the legume-Rhizobium symbiosis. *J Exp Bot*, Vol. 57, No. 8, pp. 1769-1776

'enmetsa, R. V., Frugoli, J. A., Smith, L. S., Long, S. R. & Cook, D. R. (2003). Dual genetic pathways controlling nodule number in Medicago truncatula. *Plant Physiol*, Vol. 131, No. 3, pp. 998-1008

'erez Guerra, J. C., Coussens, G., De Keyser, A., De Rycke, R., De Bodt, S., Van De Velde, W., Goormachtig, S. & Holsters, M. (2010). Comparison of developmental and stress-induced nodule senescence in Medicago truncatula. *Plant Physiol*, Vol. 152, No. 3, pp. 1574-1584

'erret, X., Staehelin, C. & Broughton, W. J. (2000). Molecular basis of symbiotic promiscuity. *Microbiol Mol Biol Rev*, Vol. 64, No. 1, pp. 180-201

'essi, G., Ahrens, C. H., Rehrauer, H., Lindemann, A., Hauser, F., Fischer, H. M. & Hennecke, H. (2007). Genome-wide transcript analysis of Bradyrhizobium japonicum bacteroids in soybean root nodules. *Mol Plant Microbe Interact*, Vol. 20, No. 11, pp. 1353-1363

'feiffer, N. E., Torres, C. M. & Wagner, F. W. (1983). Proteolytic Activity in Soybean Root Nodules : Activity in Host Cell Cytosol and Bacteroids throughout Physiological Development and Senescence. *Plant Physiol*, Vol. 71, No. 4, pp. 797-802

'ladys, D. & Rigaud, J. (1985). Senescence in French-bean nodules: Occurrence of different proteolytic activities. *Physiologia Plantarum*, Vol. 63, No. 1, pp. 43

'ladys, D., Dimitrijevic, L. & Rigaud, J. (1991). Localization of a protease in protoplast preparations in infected cells of French bean nodules. *Plant Physiol*, Vol. 97, No. 3, pp. 1174-1180

'ladys, D. & Vance, C. P. (1993). Proteolysis during Development and Senescence of Effective and Plant Gene-Controlled Ineffective Alfalfa Nódules. *Plant Physiol*, Vol. 103, No. 2, pp. 379-384

'rell, J. & Poole, P. (2006). Metabolic changes of rhizobia in legume nodules. *Trends Microbiol* Vol. 14, No. 4, pp. 161-168

'rell, J., White, J. P., Bourdes, A., Bunnewell, S., Bongaerts, R. J. & Poole, P. S. (2009). Legumes regulate Rhizobium bacteroid development and persistence by the supply of branched-chain amino acids. *Proc Natl Acad Sci USA*,Vol. 106, No. 30, pp. 12477-12482

'uppo, A., Rigaud, J. & Job, D. (1981). Role of superoxide anion in leghemoglobin autoxidation. *Plant Science Letters*, Vol. 22, No. 4, pp. 353

Puppo, A. & Halliwell, B. (1988). Formation of hydroxyl radicals from hydrogen peroxide in the presence of iron. Is haemoglobin a biological Fenton reagent? *Biochem J*, Vol 249, No. 1, pp. 185-190

Puppo, A., Groten, K., Bastian, F., Carzaniga, R., Soussi, M., Lucas, M. M., de Felipe, M. R. Harrison, J., Vanacker, H. & Foyer, C. H. (2005). Legume nodule senescence: roles for redox and hormone signalling in the orchestration of the natural aging process *New Phytol*, Vol. 165, No. 3, pp. 683-701

Raghavendra, A. S., Gonugunta, V. K., Christmann, A. & Grill, E. (2010). ABA perception and signalling. *Trends Plant Sci*, Vol. 15, No. 7, pp. 395-401

Redondo, F. J., de la Pena, T. C., Morcillo, C. N., Lucas, M. M. & Pueyo, J. J. (2009) Overexpression of flavodoxin in bacteroids induces changes in antioxidant metabolism leading to delayed senescence and starch accumulation in alfalfa root nodules. *Plant Physiol*,Vol. 149, No. 2, pp. 1166-1178

Reinbothe, C., Springer, A., Samol, I. & Reinbothe, S. (2009). Plant oxylipins: role of jasmonic acid during programmed cell death, defence and leaf senescence. *Febs J*, Vol. 276, No. 17, pp. 4666-4681

Reyrat, J. M., David, M., Blonski, C., Boistard, P. & Batut, J. (1993). Oxygen-regulated in vitro transcription of Rhizobium meliloti nifA and fixK genes. *J Bacteriol*, Vol. 175, No 21, pp. 6867-6872

Rubio, M. C., Gonzalez, E. M., Minchin, F. R., Webb, K. J., Arrese-Igor, C., Ramos, J. & Becana, M. (2002). Effects of water stress on antioxidant enzymes of leaves and nodules of transgenic alfalfa overexpressing superoxide dismutases. *Physiol Plant*, Vol. 115, No. 4, pp. 531-540

Saeki, K. (2011). Rhizobial measures to evade host defense strategies and endogenous threats to persistent symbiotic nitrogen fixation: a focus on two legume-rhizobium model systems. *Cell Mol Life Sci*, Vol. 68, No. 8, pp. 1327-1339

Sanchez, D. H., Pieckenstain, F. L., Escaray, F., Erban, A., Kraemer, U., Udvardi, M. K. & Kopka, J. (2011). Comparative ionomics and metabolomics in extremophile and glycophytic Lotus species under salt stress challenge the metabolic pre-adaptation hypothesis. *Plant Cell Environ*, Vol. 34, No. 4, pp. 605-617

Santos, R., Herouart, D., Puppo, A. & Touati, D. (2000). Critical protective role of bacterial superoxide dismutase in rhizobium-legume symbiosis. *Molecular microbiology*, Vol 38, No. 4, pp. 750-759

Sato, S., Nakamura, Y., Kaneko, T., Asamizu, E., Kato, T., Nakao, M., Sasamoto, S., Watanabe, A., Ono, A., Kawashima, K., Fujishiro, T., Katoh, M., Kohara, M., Kishida, Y., Minami, C., Nakayama, S., Nakazaki, N., Shimizu, Y., Shinpo, S., Takahashi, C., Wada, T., Yamada, M., Ohmido, N., Hayashi, M., Fukui, K., Baba, T., Nakamichi, T., Mori, H. & Tabata, S. (2008). Genome structure of the legume, Lotus japonicus. *DNA Res*,Vol. 15, No. 4, pp. 227-239

Schauser, L., Udvardi, M., Tabata, S. & Stougaard, J. (2008).Legume genomics relevant to N2 fixation. In *Nitrogen-fixing Leguminous Symbioses*, S. J. James EK, Dilworth WE, Newton WE, pp. 211-233, Springer, the Netherlands

Schell, M. A. (1993). Molecular biology of the LysR family of transcriptional regulators. *Annu Rev Microbiol*, Vol. 47, pp. 597-626

Schmutz, J., Cannon, S. B., Schlueter, J., Ma, J., Mitros, T., Nelson, W., Hyten, D. L., Song, Q., Thelen, J. J., Cheng, J., Xu, D., Hellsten, U., May, G. D., Yu, Y., Sakurai, T.,

Umezawa, T., Bhattacharyya, M. K., Sandhu, D., Valliyodan, B., Lindquist, E., Peto, M., Grant, D., Shu, S., Goodstein, D., Barry, K., Futrell-Griggs, M., Abernathy, B., Du, J., Tian, Z., Zhu, L., Gill, N., Joshi, T., Libault, M., Sethuraman, A., Zhang, X. C., Shinozaki, K., Nguyen, H. T., Wing, R. A., Cregan, P., Specht, J., Grimwood, J., Rokhsar, D., Stacey, G., Shoemaker, R. C. & Jackson, S. A. (2010). Genome sequence of the palaeopolyploid soybean. *Nature*, Vol. 463, No. 7278, pp. 178-183

Schneiker-Bekel, S., Wibberg, D., Bekel, T., Blom, J., Linke, B., Neuweger, H., Stiens, M., Vorholter, F. J., Weidner, S., Goesmann, A., Puhler, A. & Schluter, A. (2011). The complete genome sequence of the dominant Sinorhizobium meliloti field isolate SM11 extends the S. meliloti pan-genome. *J Biotechnol*, Vol. 155, No. 1, pp. 20-33

Searle, I. R., Men, A. E., Laniya, T. S., Buzas, D. M., Iturbe-Ormaetxe, I., Carroll, B. J. & Gresshoff, P. M. (2003). Long-distance signaling in nodulation directed by a CLAVATA1-like receptor kinase. *Science*, Vol. 299, No. 5603, pp. 109-112

Serraj, R. & Sinclair, T. R. (1996). Inhibition of nitrogenase activity and oxygen permeability by water deficit. *J Exp Bot*, Vol. 47, No., pp. 1067-1073

Serraj, R., Vadez, V. V., Denison, R. F. & Sinclair, T. R. (1999). Involvement of ureides in nitrogen fixation inhibition in soybean. *Plant Physiol*, Vol. 119, No. 1, pp. 289-296

Sigaud, S., Becquet, V., Frendo, P., Puppo, A. & Herouart, D. (1999). Differential regulation of two divergent Sinorhizobium meliloti genes for HPII-like catalases during free-living growth and protective role of both catalases during symbiosis. *J Bacteriol*, Vol. 181, No. 8, pp. 2634-2639

Sobrevals, L., Muller, P., Fabra, A. & Castro, S. (2006). Role of glutathione in the growth of Bradyrhizobium sp. (peanut microsymbiont) under different environmental stresses and in symbiosis with the host plant. *Canadian journal of microbiology*, Vol. 52, No. 7, pp. 609-616

Soussi, M., Ocana, A. & Lluch, C. (1998). Effects of salt stress on growth, photosynthesis and nitrogen fixation in chick-pea (Cicer arietinum L.). *J Ex Bot*, Vol. 49, No. 325, pp. 1329-1337

Streeter, J. & Wong, P. P. (1988). Inhibition of legume nodule formation and N2 fixation by nitrate. *Critical Reviews in Plant Sciences*, Vol. 7, No. 1, pp. 1-23

Swaraj, K., Dhandi, S. & Sheokand, S. (1995). Relationship between defense mechanism against activated oxygen species and nodule functioning with progress in plant and nodule development in Cajanus cajan L. Millsp. *Plant Sci*, Vol. 112, No. 1, pp. 65

Swaraj, K. & Bishnoi, N. R. (1999). Effect of salt stress on nodulation and nitrogen fixation in legumes. *Indian J Exp Biol*, Vol. 37, No. 9, pp. 843-848

Thomas, S. G., Phillips, A. L. & Hedden, P. (1999). Molecular cloning and functional expression of gibberellin 2- oxidases, multifunctional enzymes involved in gibberellin deactivation. *Proc Natl Acad Sci U S A*, Vol. 96, No. 8, pp. 4698-4703

Timmers, A. C., Soupene, E., Auriac, M. C., de Billy, F., Vasse, J., Boistard, P. & Truchet, G. (2000). Saprophytic intracellular rhizobia in alfalfa nodules. *Mol Plant Microbe Interact*, Vol. 13, No. 11, pp. 1204-1213

Torres-Quesada, O., Oruezabal, R. I., Peregrina, A., Jofre, E., Lloret, J., Rivilla, R., Toro, N. & Jimenez-Zurdo, J. I. (2010). The Sinorhizobium meliloti RNA chaperone Hfq influences central carbon metabolism and the symbiotic interaction with alfalfa. *BMC microbiology*, Vol. 10, pp. 71

Van de Velde, W., Guerra, J. C., De Keyser, A., De Rycke, R., Rombauts, S., Maunoury, N., Mergaert, P., Kondorosi, E., Holsters, M. & Goormachtig, S. (2006). Aging in legume symbiosis. A molecular view on nodule senescence in Medicago truncatula. *Plant Physiol*, Vol. 141, No. 2, pp. 711-720

Van de Velde, W., Zehirov, G., Szatmari, A., Debreczeny, M., Ishihara, H., Kevei, Z., Farkas, A., Mikulass, K., Nagy, A., Tiricz, H., Satiat-Jeunemaitre, B., Alunni, B., Bourge, M., Kucho, K., Abe, M., Kereszt, A., Maroti, G., Uchiumi, T., Kondorosi, E. & Mergaert, P. (2010). Plant peptides govern terminal differentiation of bacteria in symbiosis. *Science*, Vol. 327, No. 5969, pp. 1122-1126

van Heerden, P. D., Kiddle, G., Pellny, T. K., Mokwala, P. W., Jordaan, A., Strauss, A. J., de Beer, M., Schluter, U., Kunert, K. J. & Foyer, C. H. (2008). Roles for the regulation of respiration and the oxygen diffusion barrier in soybean in the protection of symbiotic nitrogen fixation from chilling -induced inhibition and shoots from premature senescence. *Plant Physiol*, Vol. 148, No. 1, pp. 316-327

Vasse, J., de Billy, F., Camut, S. & Truchet, G. (1990). Correlation between ultrastructural differentiation of bacteroids and nitrogen fixation in alfalfa nodules. *J Bacteriol*, Vol. 172, No. 8, pp. 4295-4306

Vauclare, P., Bligny, R., Gout, E., De Meuron, V. & Widmer, F. (2010). Metabolic and structural rearrangement during dark-induced autophagy in soybean (Glycine max L.) nodules: an electron microscopy and 31P and 13C nuclear magnetic resonance study. *Planta*, Vol. 231, No. 6, pp. 1495-1504

Vessey, J. K., Walsh, K. B. & Layzell, D. B. (1988). Oxygen limitation of nitrogen fixation in stem-girdled and nitrate-treated soybean. *Physiol Plant*, Vol. 73, No., pp. 113-121

Walsh, K. B. & Layzell, D. B. (1986). Carbon and nitrogen assimilation and partitioning in soybeans exposed to low root temperatures. *Plant Physiol*, Vol. 80, No. 1, pp. 249-255

Webb, C. J., Chan-Weiher, C. & Johnson, D. A. (2008). Isolation of a novel family of genes related to 2-oxoglutarate-dependent dioxygenases from soybean and analysis of their expression during root nodule senescence. *J Plant Physiol*, Vol. 165, No. 16, pp. 1736-1744

White, J., Prell, J., James, E. K. & Poole, P. (2007). Nutrient sharing between symbionts. *Plant Physiol*, Vol. 144, No. 2, pp. 604-614

Zurbriggen, M. D., Tognetti, V. B., Fillat, M. F., Hajirezaei, M. R., Valle, E. M. & Carrillo, N. (2008). Combating stress with flavodoxin: a promising route for crop improvement. *Trends Biotechnol*, Vol. 26, No. 10, pp. 531-537

Permissions

The contributors of this book come from diverse backgrounds, making this book a truly international effort. This book will bring forth new frontiers with its revolutionizing research information and detailed analysis of the nascent developments around the world.

We would like to thank Dr. Tetsuji Nagata, for lending his expertise to make the book truly unique. He has played a crucial role in the development of this book. Without his invaluable contribution this book wouldn't have been possible. He has made vital efforts to compile up to date information on the varied aspects of this subject to make this book a valuable addition to the collection of many professionals and students.

This book was conceptualized with the vision of imparting up-to-date information and advanced data in this field. To ensure the same, a matchless editorial board was set up. Every individual on the board went through rigorous rounds of assessment to prove their worth. After which they invested a large part of their time researching and compiling the most relevant data for our readers. Conferences and sessions were held from time to time between the editorial board and the contributing authors to present the data in the most comprehensible form. The editorial team has worked tirelessly to provide valuable and valid information to help people across the globe.

Every chapter published in this book has been scrutinized by our experts. Their significance has been extensively debated. The topics covered herein carry significant findings which will fuel the growth of the discipline. They may even be implemented as practical applications or may be referred to as a beginning point for another development. Chapters in this book were first published by InTech; hereby published with permission under the Creative Commons Attribution License or equivalent.

The editorial board has been involved in producing this book since its inception. They have spent rigorous hours researching and exploring the diverse topics which have resulted in the successful publishing of this book. They have passed on their knowledge of decades through this book. To expedite this challenging task, the publisher supported the team at every step. A small team of assistant editors was also appointed to further simplify the editing procedure and attain best results for the readers.

Our editorial team has been hand-picked from every corner of the world. Their multi-ethnicity adds dynamic inputs to the discussions which result in innovative outcomes. These outcomes are then further discussed with the researchers and contributors who give their valuable feedback and opinion regarding the same. The feedback is then collaborated with the researches and they are edited in a comprehensive manner to aid the understanding of the subject.

Apart from the editorial board, the designing team has also invested a significant amount of their time in understanding the subject and creating the most relevant covers. They scrutinized every image to scout for the most suitable representation of the subject and create an appropriate cover for the book.

The publishing team has been involved in this book since its early stages. They were actively engaged in every process, be it collecting the data, connecting with the contributors or procuring relevant information. The team has been an ardent support to the editorial, designing and production team. Their endless efforts to recruit the best for this project, has resulted in the accomplishment of this book. They are a veteran in the field of academics and their pool of knowledge is as vast as their experience in printing. Their expertise and guidance has proved useful at every step. Their uncompromising quality standards have made this book an exceptional effort. Their encouragement from time to time has been an inspiration for everyone.

The publisher and the editorial board hope that this book will prove to be a valuable piece of knowledge for researchers, students, practitioners and scholars across the globe.

List of Contributors

Hafsi Miloud and Guendouz Ali
Laboratory of Improvement and Development of Livestock and Crop Production, Department of Agronomy, Faculty of Natural Sciences and Life, Ferhat ABBAS University, Setif, Algeria

Kieron D. Edwards, Matt Humphry and Juan Pablo Sanchez-Tamburrino
Advanced Technologies (Cambridge) Ltd., UK

Jean-François Rontani
Laboratory of Microbiology, Geochemistry and Marine Ecology (UMR 6117), Center of Oceanology of Marseille, Aix-Marseille University, Campus of Luminy, Marseille, France

Eloísa Agüera, Purificación Cabello, Lourdes de la Mata, Estefanía Molina and Purificación de la Haba
Department of Botany, Ecology and Plant Physiology, University of Córdoba, Spain

Ulrike Zentgraf, Petra Zimmermann and Anja Smykowski
ZMBP, University of Tübingen, Germany

Paula Fernandez, Sebastián Moschen, Norma Paniego and Ruth A. Heinz
Biotechnology Institute - CICVyA- INTA Castelar, Argentina

David Delmail
University of Rennes 1, Lab. of Pharmacognosy & Mycology,UMR CNRS 6226 SCR/PNSCM, Rennes, France
University of Limoges, Lab. of Botany & Cryptogamy, GRESE EA 4330, Limoges, France

Pascal Labrousse
University of Limoges, Lab. of Botany & Cryptogamy, GRESE EA 4330, Limoges, France

Laurence Dupont, Geneviève Alloing, Olivier Pierre, Sarra El Msehli, Julie Hopkins, Didier Hérouart and Pierre Frendo
UMR "Biotic Interactions and Plant Health" INRA 1301-CNRS 6243, University of Nice-Sophia Antipolis, F-06903 Sophia-Antipolis Cedex, France
Laboratory of Plant Physiology, Science University, Tunis, Tunisia

Printed in the USA
CPSIA information can be obtained
at www.ICGtesting.com
JSHW011348221024
72173JS00003B/234

9 781632 395580